The Plot

*A Brilliant and Startling Theory
on the Origin and Demise of
the Dinosaurs*

By Wellington Aguiar

TEACH Services, Inc.
P U B L I S H I N G
www.TEACHServices.com • (800) 367-1844

Copyright © 2017 Wellington Aguiar
Copyright © 2017 TEACH Services, Inc.
ISBN-13: 978-1-57258-604-8 (Paperback)
ISBN-13: 978-1-4796-0854-6 (ePub)
ISBN-13: 978-1-4796-0856-0 (Mobi)
Library of Congress Control Number: 2009941126

TEACH Services, Inc.
P U B L I S H I N G
www.TEACHServices.com • (800) 367-1844

Table of Contents

Acknowledgements

Contrarily to what happens to most of the writers of their first book, I didn't have many difficulties to carry out the writing of *The Plot* bringing it to completion. It wasn't too difficult because God, the source of all wisdom and true knowledge, was with me all along the way opening avenues and providing the contacts with the right people, making it possible that this project could be brought to its utmost degree of success. I thank Dr. Richard Brown and Edward Rivera for the invaluable assistance given to me at Atlantic Union College in the organization and realization of the book launch project.

When I started writing this book, Pastor Edmilson Cardoso and Myron Samuel were of great help as they prayed with me and encouraged me to proceed with the writing process. I also thank my wife and children whose comprehension and acceptance of my absence, while involved in my writing work, greatly surpassed my expectations.

Among those who supported me with their prayers are Ricardo Veloso and wife, Felipe and Cida Nery, and other friends whose list of names is greater than the space available in the pages of this book.

I also give special thanks to my family and my wife's for the invaluable help they gave me, since all I am is in some way related to them.

Finally I thank the Framingham SDA Brazilian Church for the unconditional support given to me and to this book, and especially to my God who inspired me to write it.

Chapter One
In the Beginning

*How art thou fallen from heaven, O Lucifer, son of the morning!
How art thou cut down to the ground, which didst weaken the
nations! For thou hast said in thine heart, I will ascend into heaven,
I will exalt my throne above the stars of God; I will sit also upon the
mount of the congregation, in the sides of the north; I will ascend
above the heights of the clouds; I will be like the most High.
(Isaiah 14:12-14)*

Richard Campbell, a young man of 21, had a very active day
in the office and at school. Now that he had just gotten home for
a well-deserved rest, and was parking his car in the garage near
his home, a small suburban house where he had been living since
he was 8 years old in company of his parents and a younger sister;
his mother, Judith Campbell, a woman in her early 50s, came to
welcome him at his arrival.

"Richard, it's so good that you're home! Dinner is ready, and
I was just wondering if you would make it home on time this eve-

ning. I was worried, since there have been so many car accidents lately," said his mother tenderly.

"Mom, you know that worrying will never do any good; and since we have placed our lives in God's hands, He will take care of us. We don't have anything to fear because God is in control," said Richard trying to relieve his mother's fears.

"Yes," she said, "you're right, I shouldn't be so worried."

Richard got out of the car, picked up his books and briefcase; and on his way entering the house he kissed his mother on the cheek and said, "How are you doing, mom?"

"I'm fine. I'm just a little worried, you know," she replied.

"That's ok, mom," he said.

Going straight to his bedroom aiming to take a shower and relax a little bit before dinner, he noticed a newspaper that was sitting on the parlor table and grabbed it to get the day's news; and as he entered his room, he saw the headline on the first page: "Scientists finally complete the mapping and sequencing of the mouse's genome." "Wow, that's interesting!" he exclaimed talking to himself.

He read the whole article and started wondering how useful that accomplishment could be. The mouse's genome, being very similar to that of humans (99%), can be used very effectively to make comparisons so we can understand better our own. How many incurable diseases could be avoided if prevention could be made in the DNA level even before the child is born? The possibilities are so many that it's difficult to think of all of them now. Only the future will reveal how much significance this achievement has for all of us, since more and more ideas will be brought up as time goes by.

We have been striving to find the cure to so many diseases that have afflicted people since the beginning of the human race on our planet; and now we finally seem to be on the right track to solve much of the mystery involving these diseases. The cracking of the DNA code, which can be the solution to many of the difficult health problems that humankind faces now, can also be used to cause us harm, to greatly increase our legal problems related to human rights; resulting maybe, in a new definition of what it means to be a human being – or even destroy us.

Today the scientists are able to analyze the DNA structure and, by genetic manipulation, induce changes in its code that will benefit us with better health, better food, and maybe make our children more intelligent. But in the past there was a mind (still there is) capable of doing these things in a level that we can never imagine; and if we do, we think of it just as a fiction, something that was never possible.

When Lucifer, the covering cherub, was created; he was meant to be the most elevated being among the children of God. His intelligence was perfect and nobody was gifted with a mind like his. He had, only inferior to the Trinity, the most powerful intellect that no other created being could ever have. On this wise he was just perfect. The other angels, his partners, were delighted in hearing his wisdom and sound reasoning. They loved him and in him they placed their sympathy and confidence.

As somebody once said: "The higher the privilege, the greater the responsibility and endangerment." And that was the condition of Lucifer's life. He was created and endowed with capabilities that even today, when so much has been said about him, we can't evaluate completely how much he can achieve. We get just a glimpse of it when we consider man's capabilities in comparison to his, since we know that an angel as a superior being (a cherub, in his case) can achieve more than we ever can.

This guy, for inexplicable reasons, rebelled against his benefactor, the One who had given him life and brought him into existence. And in doing this, he could no longer keep his position before God. Somebody else had to replace him because now, being a dissident, he was trying his best to overthrow God's throne and set himself in His place as the ruler of the universe.

That was the time when the great controversy started (Lucifer's war against God); and man in the person of Adam, being involved, became an *easy* target for his attacks to revenge upon God and try to assure for himself a place from where he could carry on his battle against the kingdom of the Almighty. Since his banishment from the heavenly courts, he's been involved in warfare against the creator of the universe trying to prove his point; and parallel to all that he tries to weaken, to deceive, and to lead us astray alienating us from the source of life. The way he does it is the dirtiest way that you can ever imagine. The higher the intellect, the greater the capability of doing evil; and this concept applies to him in its full extent.

In the antediluvian world, the climax of his criminal intents was achieved when, using his powerful mind, he tried by genetic manipulation (altering the DNA code) to develop the most effective killing machines that we can ever think of, the dinosaurs.

Nowadays, after having achieved genome sequencing and mapping of human and other animal's DNA, scientists understand that the code of life, as the DNA code is known, can be effectively manipulated and by clever work of biological engineering specific anatomical and organic features can be developed in a species. Pre-established characteristics such as speed, posture (capability to stand on two legs), huge size, a voracious appetite, and a way of reproduction (laying many eggs at a time) can be obtained by developing in them both warm-blooded and cold-blooded characteristics, resembling mammals and reptiles at the same time. If these things are possible to be achieved today or, most probably, in a near future, consider how much could be achieved in the beginning by a powerful mind as Lucifer's, since he's been around even before we were created.

At this point of his thoughts, Richard was interrupted by his mother's call: "Richard, dinner is ready and everybody is waiting for you; could you, please, hurry up!"

"Ok mom, I'm on my way. Give me just five minutes to take my shower and I'll be right over," answered Richard as he ran to the bathroom.

It took Richard just a few minutes to take his shower and get to the dining room, a tiny and cozy place, where the other family members were already waiting for him.

"Sorry, folks!" he exclaimed apologizing. "I got so involved in my thoughts that I forgot that everyone was waiting for me."

"That's ok," replied his father, George Campbell, trying to help his son out of that embarrassing situation.

"Richard, have a seat so we can say grace before we dine," said Mrs. Campbell.

"Thanks, Mom," he said.

Maryanne Campbell, Richard's eighteen-year-old sister, who had spent two days at her Aunt Millie's house, was back home now and glad to see her brother for whom she had a profound respect and admiration. He was, in a certain way, her idol being, in her understanding, the paradigm of intelligence and wit.

"What has made you so deeply absorbed in your thoughts, Richard?" asked Maryanne showing a bit of concern.

"Well, children, I think we should interrupt this conversation just for a brief moment so we can say a word of prayer," suggested Mr. Campbell before Richard could reply.

They prayed; and as they finished, Maryanne brought up the same question, resuming the conversation, "Richard, what's bothering you so much?

"A headline in today's newspaper, haven't you seen it yet?" asked Richard.

"The one talking about the mouse's DNA mapping, you mean?" she inquired.

"Yes, that one," replied her brother. "Now that they've completed the mouse's genome sequencing and DNA mapping, they're making comparisons between the human genome and the mouse's. Do you know what they've found out? They discovered that there's a great resemblance between them. The human genome and the mouse's are about 99% similar (the number of human genes without a clear mouse counterpart is not larger them 1% of the total), and that we have even the gene that codes for the tail being the only reason why we don't have it, the fact that the gene, for some reason, is turned off," commented Richard having a funny expression on his face.

All three of Richard's family members laughed, when they heard that, and Maryanne said, "Imagine, I, Maryanne Campbell, one of the most beautiful girls in town having to bear the shame of carrying along a tail like one of the most repugnant of the mammals! I don't like this idea at all," she protested making fun of the situation.

"And that is not all," said Richard resuming the conversation, "with the human genome sequencing and mapping the scientists will be able to locate each one of the estimated 25,000 genes that comprise our genome; know what they code for, and when necessary, act on the DNA level to cure diseases, to develop clones, and if God doesn't intervene, try to shape the future generations according to the purposes of evil minds that may eventually come into the scene.

"But what concerns me most is not the human misuse of this wonderful technology now available. What makes me deeply worried is the fact that there's a being (and not everybody seems to be aware of it) who's been around for a long time, even before us, who knows all this in a deeper sense than we can ever know, and used it in the past to carry out his malevolent plans to eliminate humankind. He developed the great variety of dinosaurs. The huge, medium and small sized herbivores, and did the same with the carnivores making them into strong, and in some cases, indestructible killing machines."

Maryanne was astonished when Richard finished speaking. She was so impressed that she couldn't even utter a word. Then, at this point, their mom interrupted the conversation and said, "Richard, I love your reasoning and I believe that something like that might have happened; but I have to tell you that if you don't stop this conversation right now and eat, you are going to be a serious candidate to die sooner than you think, and the bad guy you've been talking about won't need to use dinosaurs to kill you," she said trying to put some humor in her words.

And as she said that, everybody laughed.

Chapter Two
Meet Two Pals

In the following day, Richard got up early as usual and hurried up to the university where he would attend a metallography class. As soon as he entered the corridor leading to his classroom, he met his best friend Paul.

Paul Silvers was Richard's best friend, a 23 year old, blond-haired guy, with a short haircut whose face seemed always to be illuminated by a sincere smile. Richard and Paul had known each other for years, and Richard considered Paul more like an older brother. Paul's special interest, besides his engineering course, was unsolved mysteries. He used to watch all those mystery shows on TV; but though being a huge mystery fan, he wasn't the superficial type, as might appear at first sight. He was an intelligent and cultured guy; and it could be said that he was an individual whose mind was beyond his years. He was capable of listening to people and paying attention to what they were saying even though their ideas, most of the time, didn't agree with his personal beliefs.

"How are you doing, Richard," said Paul as he approached his younger pal.

"Pretty good, how about yourself, Paul?" asked Richard.

"I'm fine, thanks," answered the friend. "We're a little bit early for the metallography class; would you like to go to the cafeteria so we can talk a little bit?

I have some things that I want to tell you," he added.

"What's that?" Richard asked.

"It's about an invitation I've got for a lecture that is going to be given on cloning and its implications on society," explained Paul.

"When is it going to be held?" asked Richard, curious.

"Tonight at eight o'clock, at the medical school, where Martha, my girlfriend has classes. If you don't mind, I can pick you up at seven, since you don't know the way," offered Paul.

"Ok, Paul. Pick me up at seven," agreed Richard.

Richard was especially excited about that lecture, since genetics, DNA and cloning where subjects somewhat related to the issue he had been focusing his attention on for a long time – dinosaurs. He was strongly inclined to believe that "the terrible lizards", as they were called by Richard Owen, the man who studied them first and gave them the name, had a strong connection with modern research in genetics. In other words, he thought that those huge animals had come into existence not by evolution, but as a result of a well planned and executed conspiracy against the human race.

*** *** ***

Doctor John Watkins started his lecture explaining how biology had evolved along the years, the hardships faced by those who dedicated their lives to its study, some of them being not even recognized in their lifetime. He gave the audience a briefing on the main discoveries during these nearly 200 years of study of the subject starting with Gregor Mendel's research in the mid-1800's, Watson and Crick's discovery of the structure of DNA in 1953 that made it possible for other scientists to crack the "The Code of Life" bringing it to what is starting now to configure as the "Age of Genetics".

He also explained about the four nucleotide bases (A, T, C and G) that form the DNA strand which when lined up in a certain sequence, they form the genes responsible for heredity or even the determination of certain features that characterize the different life forms and species living on the planet.

He talked about human cloning and explained why, so far, it's difficult (or maybe impossible) to successfully clone a human being using the techniques now available, since the mature cells of the human body have many of their genes switched off, and in order to revert them back to the stage when the cell is still undifferentiated (in this stage the cell can grow into any other type of cell in the body such as skin cells, hair cells or brain cells, depending on the set of genes that are switched on in that specific cell) they have to use a process that might cause the embryo to have flaws in its development.

Many of the scientists believe that cloning human beings is not safe, there having other issues to be addressed, such as the laws which will regulate it, the cloned person's rights, eventual health problems decurrent of the process, and many other issues that we cannot even think about now.

The lecture was a long one, and lasted for about two hours. When it was finished, Richard felt completely satisfied because Dr. Watkins had touched some important points that he didn't understand clearly as yet. For example, Richard didn't know for sure if it's possible to develop according to a pre-established project, a new species having some anatomic and organic characteristics previously determined. When the lecture was finished, Dr. Watkins gave to the audience the opportunity to ask questions and Richard didn't miss the chance to ask for explanation on that particular point.

Richard's question was answered as follows: "Nobody has ever done it, though we believe that something like this might be possible to be achieved in the future. Today we're struggling to develop a cloning technology that can make it possible for us to clone human beings. In the near future manipulation of the DNA strand might give us the ability to develop certain desired characteristics or features in an animal or plant.

'Since all the nucleotides (A, C, T, G) found in the DNA molecule are like a mini-alphabet that spells out messages for the cell, different combinations of these letters spell out codes, and these codes tell the cell which specific proteins it should make. Proteins are very important for all the workings of a cell and, in essence, control a great deal of what happens within a living organism. Proteins are also responsible for many of the traits of living things. So the DNA molecule is responsible for passing hereditary information by instructing the cell to make proteins that will influence the growth, development, the appearance, the total anatomy and the fact that the living thing will be a plant, an animal or a human being' (Decoding Your Genes). And of course, the intelligent manipulation of the DNA code will determine the desired changes. These things are not likely to be achieved now, but in the future it may be perfectly possible for us to perform successfully this kind of experiment. This is what I would call the development of a new species effectively using the Code of Life, the language of creation."

When Richard, Paul and Maryanne left the lecture hall, Richard was deeply convinced that his theories were not absurd; and since we, limited human beings, got to this point now, somebody whose mind is extremely superior to ours could easily have done this in the past. While they were still walking to the car, Paul asked, "How did you like it, Richard?"

"Well, I liked it very much," answered his friend. "It gave me the opportunity to clarify some of my ideas about some other issues, and the understanding that all this knowledge on the DNA structure, the genes and their function in the cell determining characteristics and anatomic features (for which they have been programmed) in all living things, will give the scientists the ability to map the human genome, to locate all our genes, find out what they code for, and finally lead them to the cure of many deadly diseases that have taken the life of millions of people," he added.

When Richard finished his comments, the three of them were already tucked into the car heading toward the Campbells' house.

Maryanne looking out the window saw the cloudless, dark, starry sky. She wondered how beautiful and complex the universe is; and as if in a moment of ecstasy she asked her brother, "Richard, don't you think that we, human beings, though small and apparently insignificant when compared to the greatness of the universe, are in a certain way more complex and important than all these galaxies and worlds that we have heard about?"

"Of course I do," answered her brother.

"And why do you think so, Richard?" she asked once more.

"The universe was created by the word of God, and by His power the worlds were established. He ordained and all of them came into existence," he said paraphrasing the Bible. "But regarding to man, it is said that we were created by a personal being, in His image and likeness. He stooped down and with His own hands executed His project. Using the language of creation – DNA code – He put everything in place and made sure that the necessary enzymes and proteins would be produced in order to keep life going. But even before the DNA could be able to accomplish its purpose, God breathed into man's nostrils the breath of life, which is to say that He took life (its energy) from Himself and gave it to man. It wasn't only a spark of life, but a constant flowing of it, since until now and forever there will be no life derived from any other source, but God," answered Richard as if he were in a moment of great inspiration.

"Oh, that's great," said Paul, "I like it!" he exclaimed in a burst of enthusiasm.

"You know, Paul," continued Richard, "when I consider man from the same stand point as Dr. Watkins, I see the complexity of our body and the nature of it; and in so doing I just can't admit evolution. Scientists, in their search for the truth, delve very deep into the nature and structure of human beings, but there's always a point beyond which they can't go. They know that only 3% of the genes in the human genome are actually coding material, being the remainder 97% all junk DNA that codes for nothing. Why is it there? They don't know. It's like in the scientist's attempt to explain the origin of the universe, the Big Bang; they go back peering into the past with their potent telescopes getting pretty close to the beginning, but never being able to see God in the moment of creation as the cause and origin of everything. That's why I think they can't explain convincingly the origin of life."

"That's right," said Maryanne.

"This is one of the reasons," observed Paul, "I am glad we're friends, Richard. We have a lot in common, and among the things that I love, as much as you do, is the truth in all its amplitude and multifaceted aspects. The fact that we can always talk about it, since you are one of those who are seeking to know and apply it to their personal lives, makes me feel pretty confident that a relationship like ours can never go wrong and we will always be friends."

Paul's opinion about Richard wasn't exaggerated. The younger pal was really a seeker of the truth; and in his opinion it wasn't something to be applied only to the moral aspect of our lives, but something more comprehensive reaching far beyond our limited understanding. In Richard's opinion God as the supreme manifestation of truth, is the point to which everything converges. The truth about the universe (its beginning and how it functions); the moral aspect of the truth which governs our dealings with one another; and that aspect of the truth which refers to our physical bodies (how diseases occur and how they can be cured), all of them being important parts of the whole, the universal truth.

"When Paul Silvers finished his comments, the car was already stopping at the Campbell's residence, located in a very quiet street and embellished by a small and very well tended garden filled with roses whose scent inundated the atmosphere.

"Good night, Maryanne and Richard!" said Paul.

"Good night!" said the two siblings in unison.

"I'll see you tomorrow in school," added Paul.

"See you," answered the younger friend.

Chapter Three

Magnificent Project

The next morning, when the Campbells got to church, just before the beginning of the worship service, they found most of the brethren already there. It was a beautiful sunny morning and everyone seemed to be full of gladness and optimism.

Maryanne was wearing a light blue dress that made more noticeable the contrast and beauty of her long black hair and white skin; making of her, at the age of eighteen, a graceful and delicate girl. Her mother, as usual, wore her common church clothes whose simplicity, modesty, and fitness revealed the good taste for clothes that characterized the Campbell women. Richard and Mr. Campbell wore their best suits for the occasion.

As they entered the church, the Campbells chose a pew on the forefront of the nave and there they sat. The platform was composed and the worship service had already started. The speaker in that morning had chosen the text of Isaiah 28:29, which reads: "This also cometh forth from the Lord of hosts, which is wonderful in counsel, and excellent in working."

The main line of thought in the sermon was the wisdom of the Lord's counsel; the fact that He always sees things in their right perspective, and never makes mistakes in judgment or counseling. It referred also to His comprehensive mind. He can think of everything at the same instant with extreme preciseness and accuracy.

For instance in the act of creating man, He thought of every single detail without forgetting anything, and at the same time He was able to see, in His omniscience, how it would work after being put in place performing its natural function. Each of our cells, their innermost parts – DNA, genes, and chromosomes – containing all the information about proteins and enzymes making that make it possible for life to reproduce itself in a certain way preserving all the species; all these details were already in His mind, and they didn't happen by chance; but were meticulously planned and perfectly executed. That refers to the wonderful mind of God and tells about His immensurable capability of creation; since when He finished His work, He declared that it was "very good" which means no flaws, no imperfections at all.

"His excellent working." What does that mean? That means: speed, preciseness and perfection.

In order to understand God's way of working, we should consider the way the men plan and execute their projects; how much effort it takes, the time and money involved. Let's consider, for instance, the pyramids. It is said that a tremendous amount of time and effort was required to build them. Located in the Giza plateau is the tallest pyramid of all, built by the great Egyptian Pharaoh Khufu, which stands taller than a forty-story building. It was constructed of 2,600,000 blocks of stone, weighing on average 2.5 tons each. This remains undoubtedly as one of the greatest projects that humans had ever undertaken at that time, requiring a great deal of planning, labor and money (some believe that no slave labor was involved, since the archaeological findings indicate that the workers were paid for their labor).

Now, with so many technological advances such as powerful computers and other sophisticated machines, we can achieve a lot more than the past generations could; though we still have a long way to go in order to make our planning and executing as much close to perfection as limited humans can make it; being the shortness of our lifetime one of the greatest problems that we human beings face regarding our projects. We have been lingering in the field of science for as long as six thousand years and only in the last fifty years of the twentieth century we were able to put all the pieces of human knowledge together and achieve our most advanced developments in the medical field, in the space program and in biology; coming to what is called by many "The Genetic Revolution". So much light has been shed in this area of human knowledge that genetics is seen now as one of the most promising areas for capital investment. With the discoveries of the structure of DNA, and with human genome sequencing and mapping, the identification of disease genes will be facilitated leading to the cure of many genetic disorders freeing our future generations of maladies such as breast cancer, schizophrenia, diabetes and Alzheimer's disease. The possibilities are so many that they cannot be counted without the risk of leaving something out.

At this point of the sermon, Richard began to imagine the magnitude of the achievements of God's mind compared to those of Lucifer's (since he was an angel only inferior to the Trinity) and man's mind. It didn't take Lucifer too long to develop all the knowledge necessary to figure out the DNA structure and how it works. To human beings it took about six thousand years to get to the point we are today, but Lucifer, as a superior being having a more powerful intellect, was able to arrive to these conclusions in much less time than we did. This advantage gave him the ability of developing the dinosaurs in a relatively short period of time.

That was certainly a magnificent project, and a great genius was required for its development. Imagine what it means to project a living animal (dinosaur) having in itself the most important characteristics found in other successful species thriving on the planet such as mammals, birds, and reptiles. The hip bones of birds that gave them the capability to stand on two legs; the metabolism of warm-blooded animals found in mammals (creating a great need for food in a shorter period of time); and cold-bloodedness found in reptiles whose way of reproduction by laying eggs extremely favored some of the dinosaurs that, in some cases, could lay 100 eggs at a time; the relatively high speed of locomotion (Tyrannosaurus rex could travel at 30MPH); and viciousness of the Velociraptors, which hunted in packs. Some of the herbivores such as Diplodocus and others could eat up to 450 pounds of foliage a day; and considering the size and strength of these huge plant eaters, it's not difficult to imagine the great impact they caused on nature provoking environmental disequilibrium making it difficult or even impossible for humans and other species to survive.

When the service was finished and the Campbells left church, Richard was still absorbed in his thoughts about Lucifer's magnificent project of the dinosaurs; and during their ride home, Maryanne knowing his special interest on the subject came up with the question, "Richard, do you really think that there could be somebody, besides God, capable of developing a new species like the dinosaurs?"

"Of course I do!" answered her brother. And in my opinion it was just a matter of time. God has the capability and knowledge to do it in a moment; man one day might be able to develop a knew species according to a project, but Lucifer certainly was able to do it long ago. Even though life still comes from the same source, which is God, the mastering of the code of life could make it possible to any other being (angel or man) to manipulate the DNA strand, change its sequence and consequently alter the genetic code of an animal cell obtaining as a result a totally different organism. Presumably he didn't get to the final result right away having to spend a long time doing his experiments, starting with Postosuchus, Coelophysis and finally reaching the stage of development of Tyrannosaurus rex which I believe was his masterpiece.

That animal, besides being huge, had special features that made it into a powerful killing machine. Archaeologists have found a five-foot-long T Rex skull; and considering the large opening for the olfactory organ, they concluded that it could smell its prey from a long distance; and according to some specialists it could also eat a great amount of meat at once. I don't think there was a man that could kill that beast, especially considering that there were no appropriated weapons to kill it," added Richard emphatically.

"But the scientists today have their laboratories and sophisticated equipment to do their research; do you think that Lucifer would need all this to carry out his experiments?" Maryanne asked pointing out a flaw in Richard's theory.

"One of the most sophisticated machines that we have today is the computer. It helps us to perform very long and complex calculations with exactness beyond our human capabilities. It stores tons of information and gives us the answer to most of our hard-to-answer questions depending on the program, the software and the data we feed the machine with. A mind notably more powerful than the mind of our most brilliant scientists can do very well without these machines; and regarding the laboratory, he might have had one – since he needed to perform his experiments – and have destroyed it afterwards. God doesn't need to perform experiments because He goes from the beginning of the project to its end knowing exactly what's going to happen, since He is omniscient. The fact is that just because we can't find the lab or any vestige of it, it doesn't mean that he couldn't have had one even more sophisticated than the ones we have today. Remember, we are talking about a being who is a whole lot more capable than any of us humans." explained Richard.

"Recently, I read in a book,"intervened Maryanne, who had been attentively listening to Richard's explanation, "that in reference to the cracking of the DNA code we're like kindergartners who are just learning how to read. Maybe the fact that we can read the code of life should also help us to understand and accept the truth about dinosaurs, and tell us that it's time to start looking in a different direction – not to evolution – in our search for answers concerning those huge animals. To many, this might appear to be foolishness, while to the few it certainly will be the opportunity to find out the truth. After all, concerning this matter, who can say that the majority has the truth? You know what history says about Galileo Galilei and others who had, in a certain moment of their lives, to take a stand for the truth even though the dominant opinion was the opposite of theirs."

When his sister finished talking, Richard remained silent for a few minutes. The Campbell's car was already on Main Street heading toward home; then Richard started thinking how difficult it is to disagree with the majority defending an unpopular point of view. He thought about the most beautiful truths that could benefit those who would accept them, but are set aside because it's not convenient to take a stand for them. It would cost too much to abandon their most cherished beliefs in order to embrace a new and different concept of life. Not everybody would be willing to pay the price.

Mr. Campbell, who had listened attentively Maryanne's comments, added, "Yes, you're right. Most of the people are interested in the truth, but only the part that is convenient to them. Since it encompasses all the aspects of our lives – moral and spiritual – the universe and the laws that govern it; with everything converging on God, who is the most perfect expression of truth. When it calls for a major change in ideas and life style, people don't want it."

At that point, Richard added, "Well, this has been common the whole history of mankind. It's easier to go along with the majority and change position whenever it's convenient not taking a stand in opposition to the crowd."

"As a matter of fact, Richard, what is the crowd?" asked Mrs. Campbell who, so far, had been only listening. "Let me tell you," she continued, "the crowd doesn't have a name, personality, responsibility, or commitment to the truth. They just want to do what is good for the moment, and tomorrow, if convenient, they will change position. But the truth, my dear son, for a few good people, is something to live by and to die for. They will remain on the side of truth, even if it is against them. What does history say about Abraham Lincoln? He was a man who believed in, and lived by, the truth. And this is the best principle that a person can have in life," said Mother concluding her thoughts.

While all this conversation was still going on, Mr. Campbell parked the car in the garage, the family got out, and Maryanne, jokingly, made a funny comment, "Richard, let me tell you one of the greatest truths about myself at this moment. I'm truly hungry and if I don't eat in the next few minutes, I'll be in serious trouble and unable to speak about the truth with anybody." As she said that, she ran inside the house and Richard having a grin in his face reminded her, "Don't forget, Maryanne, 'Man shall not live by bread alone...'"

Chapter Four

A Better Reality

> *Most of the change we think we see in life,*
> *Is due to truths being in and out of favor.*
> Robert Frost (1874-1963)

It was approximately 11 PM. Even though Richard was already in his bedroom, he could not fall asleep. He was very tired after a very active day, with church in the morning, followed by visiting with friends all day long. It was always like that, because he usually spent most of his weekly day of rest and worship going from place to place and doing volunteer work.

Richard was absorbed in his thoughts when somebody knocked on his door. His sister's voice came from outside: "Richard, it's me! Are you awaked? Let me in!" She asked affably.

"Yes," Richard answered, "give me just a few seconds, I'll let you in." And as he opened the door, he asked, "What's going on, Maryanne?"

"Nothing special," she replied. "I just couldn't sleep so I decided to come by for a little chat. We haven't talked in ages," She added with a smile on her face.

"Oh, sure, come in!" exclaimed the older sibling. "I'm in the same situation as you, my sister. I can't fall asleep either. Why don't you sit down, relax and we'll talk?"

Since her infancy, Maryanne had developed a special affection for the older brother. As a matter of fact, instead of being rivals (as often happens to siblings), they were best friends; and even in his childhood, Richard was always very protective of his younger sister. He used to say that she was "the apple of his eye;" and Maryanne always appreciated her brother's protection and care.

"Ok, beloved sister, what's going on?" he asked having a friendly expression in his face to make her feel more comfortable.

"Well, since I couldn't fall asleep, I started thinking about the sermon we heard in church this morning. It's the expression that says that God is "wonderful in counsel and excellent in working" the preacher spoke about. The wisdom of God's counsel and His effectiveness, speed and exactness in performing His work are amazing," she explained. "Having read a book on Gregor Mendel's biography and experiments to establish a theory on how genetic traits are transmitted from one generation to the next, I started wondering about how God's providence works making that these discoveries in the genetic field, meaningful and revealing as they are, may lead us to a totally different explanation on how life began in this planet," She pointed out wisely.

"As a matter of fact, surfing in the Internet, I found out that there are some scientists who advocate the idea of using the technology on DNA to trace back humankind's origin; there having at least one of them, who thinks that the evolutionary theory is no longer the best explanation for the origin of life in our planet – or in the universe – since the process of protein formation is too complicated, and is not likely to have happened by chance. According to this scientist's idea, it's unimaginable that a difficult and complex phenomenon as life didn't have an all-wise planner behind it," she concluded.

"And what is it that you read about Mendel, my dear? You've got me interested, please tell me more!" Richard urged her.

"Johann Gregor Mendel was an Austrian catholic monk living a peaceful life in a monastery in Europe," She began, "being one of his main activities to care for the garden in which peas and other vegetables were grown. After a while doing his agricultural job, he noticed that there were a lot of variation among the peas with each kind presenting different characteristics in color and in shape. He started crossing different kinds of them and observing the results; and after a while – and some very hard work – he was able to come up with the theory on the factors of *heredity*. Later, other scientists would name those factors of heredity as genes. By the time he concluded his classic breeding experiments on garden peas in 1866 and published the results, Charles Darwin, the great naturalist, was already famous in the scientific world for publishing *The Origin of the Species*, in 1859. Some historians say that Mendel sent a letter to Darwin telling him about his observations, but never getting an answer from the scientist who was by this time enjoying his reputation among the men of science.

"I personally believe," she proceeded, "that Mendel's theory was ignored by Darwin because he might not have seen how to fit it in his theory of evolution. Later, in the early 1900s, three plant breeders discovered the forgotten work of Johann Gregor Mendel; and Thomas Morgan, a respectable embryologist, also in his forties, dedicated his efforts in identifying the physical bases for hereditary factors. He performed his experiments using various different species, but the one that made him famous was the fruit fly, Drosophila.

"By the fiftieth anniversary of Mendel's famous experiments, the chromosome theory of heredity was all but established, and the scientific revolution on the field of genetics was underway," said Maryanne finishing her account of Mendel's story.

"It's very interesting, Maryanne," observed Richard, "how some facts in history fit to each other like the small pieces of a huge puzzle. Approximately by the time Mendel had finished his experiments sending that letter to Charles Darwin and other people, there was a German philosopher named Georg Wilhelm Friederich Hegel, whose 1855 theory on *dialectics* was gaining popularity in the world.

Hegel, in his dialectical concept, proposed an explanation for the precedence of the spirit (mind) over matter; and in general lines he stated that every reality (or concrete fact) has within itself the seed of its own negation (denial or contradiction), and if it is left to follow its normal course of development, it will finally deny the primeval reality's existence, and establish a new reality instead, more perfect than the first. For instance, an egg, which is a concrete thing or reality, having its degree of perfection, is in a certain way, the most perfect manifestation of that particular reality. It has a *yoke*, a *white*, a *protective membrane* and its *shell*. Over time, if the right conditions are provided (air, warmth, and protection), the substance inside the egg (the germ of its own negation) will evolve into a new reality which constitutes the chick that will appear when the egg (the former reality) is hatched and will, in its turn, deny the egg's existence as such, forming a new reality, the chick, more perfect than the egg. This philosophical concept can be applied to explain the facts in history, being considered by many as the engine that propels history; and those who advocate this principle as an expression of the truth, are said to have a *dialectical* vision or interpretation of history.

"In my opinion, which is also the opinion of many other people, there's a strong evidence of a power, which I call *providence*, underlying all the progress that is made in science. We can't see it, but if we pay attention to the order in which these progresses happen, we are going to be able to establish a relationship among them all. For instance, the genetic revolution, that began with Gregor Mendel, climaxing with the discovery of the DNA structure and the cracking of its code; may, certainly, be the seed of the negation of evolution, as we know it now, and establish a new reality more perfect in its description of facts in nature than the evolutionary theory is. The truth in its universal aspect provides an explanation for everything that happens in the universe including our own life. And the present truth being the unfolding of it (the truth) at each point in time, provides an understanding of the universe, our world and ourselves for the present time, since it's cumulative and encompasses all the knowledge and information ingathered all along human history. There has never been – neither in the present, nor in the past generations – not even a single person who possesses all the knowledge of truth at one time; but on the other hand there have been people who possess the knowledge of truth for their time (the present truth) which comprises each partial revelation of it given to us over these six thousand years of human experience in this planet.

One of the reasons why science is so credited among us is because it deals with the truth. Being all the progress in technology that makes human life more enjoyable, and the development of new medicines to fight successfully diseases responsible for the death of millions, a powerful argument on behalf of it regarding its capability to understand and manipulate the facts of truth related to the physical and biological laws that govern the universe and our own beings. Every scientific discovery that is made adds to our knowledge of the truth as a whole; and a scientific concept from the past might not be valid today, since it was based on false assumptions; and when a new insight on that aspect of the truth was brought to light, a new and better description was given to it. The truth doesn't change, but new insights can deepen and even improve our understanding of it," Richard concluded his long explanation.

When he got through with his explanation, Maryanne was deeply absorbed in her thoughts, and then as if she were awaking from a dream, she suddenly came back to reality and asked, "Richard, do you think that all this knowledge on genetics may provide a consistent basis for the negation of the theory of evolution and the origin of dinosaurs?"

"It certainly will," he answered, "but what we need to know is if there will be willingness to do it, since this dinosaur thing has become a very profitable business, and not many people would be willing to give up their profit on behalf of a new interpretation of the story more in keeping with the truth.

"At the end of the Middle Ages and beginning of Modern Age, fourteen to seventeen centuries," Richard continued his explanation, "Europe witnessed a great revolution in the fields of arts, science and philosophy changing radically our view of ourselves, religion, history and the universe called the *Renaissance*. This cultural revolution took place because the learned men of the time, considering the limitations of church dogmas on the origin of Homo sapiens, the other life-forms living on the planet, and even the origin of the universe and how it works; decided to abandon the way of thinking of the clergy and return to the classics found mainly in the Greek and Western European cultures.

Now, in order to get a better panorama of history, science, mankind and the universe, organizing the pieces of the puzzle in a more understandable fashion; we need another revolution, a modern Renaissance, taking into consideration not only the classics and modern experimental science, but also what the scriptural record says about our origin, our destiny and the origin of the universe; since trying to explain human existence and the universe's ruling out the presence and the creative work of a personal God is an impossibility.

And in view of the fact that God is the ultimate expression of truth in all its aspects; a simple fact of science, if well understood, should be able to teach us a great deal about Him; and His role as the creator and the preserver of our world. The problem that we face today is more related to the way the men of science interpret the informations painstakingly gleaned from the fields of research and fit them together in order to get a complete picture of ourselves, history, and the universe that is harmony with truth. I wouldn't go to the point of saying that there's a lack of honesty among scientists, and as a matter of fact, they must be commended for their hard work because otherwise what would become of us without the knowledge provided by the men of science of our age? What would our life be like without the medicines, vaccines, the environmental improvements provided by the scientific advancements that take care of pollution, agricultural production and other vital aspects of our daily lives? Time and again I come to the same conclusion that we need an explanation about human origin, and the origin of the universe that takes into consideration the truth in its entirety. What God says in the Bible, what nature reveals in its natural laws that operate in harmony with the Word of God; and what science says, as far as it is in harmony with the Holy Scriptures."

Chapter Five

Specialized Meat Eaters

It was a Sunday morning, and all the Campbells were at home; and since Aunt Millie had come for a visit, Mrs. Judith urged her to stay for lunch. Richard, Maryanne and Paula, Aunt Millie's 19 year-old daughter, were having a lively discussion around the grill, where Richard was preparing the barbecue for lunch. Paula was an intelligent and interesting person who always had a new idea to throw into a conversation making it even more exciting. It didn't matter if she were talking about something trivial or a very important thing; she always had a way to get people involved and interested in the subject under consideration.

When the meat was set on the grill and it started roasting, Paula caught a whiff of it and said jokingly, "I'm so hungry that I could eat all of that meat like a hungry dinosaur – all of it in just one gulp!"

Richard laughed and said, "If you knew a little bit more about those beasts, you wouldn't be saying that."

"Why? What is so bad about them?" she asked a little surprised.

"Those animals had no trace of civilization, and even the slightest sign that they could be tamed. They were the ultimate machines of destruction specially, programmed to kill," he answered.

"Well, you must know a great deal about them. Maryanne told me that. The only thing I know about dinosaurs is that some of them were carnivorous and others, herbivorous. That's all. Could you please tell me more about them?" she asked.

"Oh yes, I can do that! What do you want to know?" he asked.

"Tell me about the meat eaters, the carnivores; I've been always curious about them," she answered.

"Dr. David Norman, Ph.D., head of Paleontology at Nature Conservancy Council and Research Fellow at Oxford University, says that the name *dinosaur* comes from two Greek words: *deinos*, terrible, and *sauros*, lizard. Professor Richard Owen named them in 1841 after having analyzed some bones, concluding that they were a different species, thus deserving a different name; and he came up with the Latin name *Dinosauria* referring to the totally new species.

"Ever since dinosaur bones were first discovered, there have been people who see some relationship between the *terrible lizards* and the mythical dragons; there having some cultures like the Chinese, Greek and even the Hebrew in which the dragon is a prominent figure. In the Hebrew, the names serpent and dragon are attributed to Lucifer possibly in an attempt to relate him to his audacious work in developing the dinosaurs, his masterpiece at the time."

At this point, Paula interrupted him and said, "Are you saying that the dinosaurs and the dragons are the same creatures?"

"There seems to be a relationship between them, even though the myth talks about an animal that could breathe fire out of its mouth and we know that nobody has ever said that about dinosaurs. The legend of fire-breathing dragons could have been added some time later maybe in an attempt by Lucifer to dissociate the mythical dragon from dinosaurs," answered Richard.

"But they say that dinosaurs lived millions of years ago. How could they have existed at the same time as humans since our species has been on the planet for a period of time that can be measured by only a few thousand years?" she asked a little bit confused.

"Some scientists," Richard started, "have determined the age of our planet by the amount of certain isotopes decay. The most commonly used isotopes are Rubidium-87/Strontium-87 (where the Rubidium decays to form the Strontium), which is most often used for geological dating; and Potassium-40/Argon-40 (Potassium decaying to Argon). Measured by these isotopes rates, some of the oldest rocks have been determined to be approximately 3.7 billion years old. But talking about the planet's age and the beginning of life in it, are two totally different things; since earth itself could be that old (I don't think the scientists are wrong on this particular point), but human life and the other living things on the planet are not older than six thousand years. And the fact that there's a disagreement between science and the Bible account is due to a misinterpretation of the geological record. The scientists got the right information from the rocks, but in their eagerness to eliminate God from the story of man and the universe as the creator, they built up all these arguments around the evolutionary theory relating our existence – and that of all the other life-forms – to an evolutionary process that started from bacteria and mollusks and evolved to what we see today. This is the main reason why we don't have a clear picture of the universe's origin, and our own. The experimental science and the Bible record should be in agreement, since they speak about the same things when it comes to man's and the universe's story," Richard asserted positively.

"Since one of the greatest desires of mankind," Richard continued, "is to know the answer to the questions which refer to our origin, 'why are we here?' and 'where are we going?'; scientists have been trying to answer them and satisfy this deep longing of the human soul. In this attempt, the theory of evolution has been supplied with the most sophisticated arguments that human intelligence can ever think of; but the evolutionists, though educated and knowledgeable in science, have not been able to put forward an indisputable theory that would reduce the creationists to silence. The evolutionary theory is sophisticated, I admit, but it doesn't reflect the truth. As a matter of fact, its sophistication tells us about its origin and originator, Lucifer, the being who has been trying, from the very beginning, to blot out of our minds the idea of a personal God and a loving creator," concluded Richard.

"That's a very interesting point, Richard," Paula said, "but how about the dinosaurs?" she asked trying to bring the conversation back on track.

"In his book entitled *Dinosaur!* Dr. David Norman says that only the reptiles that lived on land are included in the category of dinosaurs and all of them walked on upright posture supported by their pillar-like legs. Their legs were arranged right beneath the body in opposition to the sprawled leg position of some other reptiles such as crocodiles. This particular arrangement of dinosaur's legs is very similar to that seen in living birds and mammals. It allowed them to swing the legs beneath the body so they could have long-striding legs and run fast. In that way the legs could better support the heavy weight of the largest dinosaurs acting literally like supporting pillars.

"Scientists have made a division of dinosaurs," Richard continued, "according to the bones of their hips: the saurischians (reptiles-hipped) and the ornithischians (bird- hipped). Among the saurischians we have the theropods which are mostly carnivorous, there having many of them; and the most known being the Allosaurus, Tyrannosaurus, Coelophysis, Compsognathus, Deinonychus and Velociraptor. The name theropod means, "beast-foot", referring to its three-toed foot, with each toe ending with a sharp bird-like claw. They had long powerful hind limbs; slender or lightly built arms; chest short and compact; body balanced by a long tail at the hip; neck sharply curved and flexible; a head equipped with large eyes; long jaws always lined up with dagger-like teeth.

"Some scientists argue if the dinosaurs were cold-blooded or warm-blooded animals; there having some who believe that in dinosaurs the two characteristics were cleverly combined making them into very specialized killers. The warm-bloodedness gave them much in common with mammals, and as such they could grow very fast. Consider, for instance, a human being at the age of 8 who is still a child; a dinosaur at this age was a fully-grown animal, some of them weighing up to 12 tons. Not long ago, they discovered a dinosaur fossil having a four-chambered heart, like the mammals; and that, according to some specialists, could give the beast the activeness found in the mammals.

A recently-found specimen of Tyrannosaurus rex whose skull measured 4ft (1.2 m) with long teeth measuring 8 inches (20cm), serrated on its back surface, has changed the opinion of scientists quite a bit; since before, the sophisticated predator was thought as a sluggish and slow-witted animal, and now the idea is that T. Rex could be very agile running up to 30mph; could have developed an acute sense of smell and a well-developed intelligence (based on the size of the brain cavity in the skull), since large brain seems to go along with more intelligence.(See appendix 1)

"According to Doctor David Norman, another carnivore that has been discovered to be a very specialized predator is Deinonychus, meaning terrible claws. Quite small in size for the standards of huge dinosaurs, measuring about 8ft (2.4m) long; having a large head; a strong neck; long grasping arms and fairly long legs. The skull of this predator being considered large for its kind is surprisingly light weighted because of its window-like openings. Having large jaws and powerful muscles to make it capable of a very strong bite and able to gulp large chunks of meat at once; these animals being not used to chew the meat and then swallow it up; were in this way enabled to eat large amount of meat in a short time. And in addition to all this, they were equipped with long arms; three-fingered hand ending in viciously curved claws; they had stereoscopic vision; that is, they could see an object with both eyes at once, allowing them to judge its distance accurately; they had large brain for reptilian standards; sharp sense of smell and a second claw specially designed to kick and slash disemboweling the prey," Richard ended his description pausing a little bit.

Maryanne, then, asked a question, "Do you think there's any kind of living predator today that could match the sophistication of Deinonychus?"

"Some scientists say that there's no predator that resembles Deinonychus at the present time; and that should tell us about its uniqueness and purpose as a specialized killer," answered Richard.

"This is one of the reasons why I just can't believe in evolution," said Maryanne, "such a perfect animal could never have evolved in the way they say it did. It has to have been planned and developed by an intelligent mind. See, for instance, the different kinds of reptiles and mammals living today; and there's no way to compare them because the dinosaurs are totally different. Predators like the lion, the tiger and others; they are not meant to destroy; their need for food doesn't make them into killing machines like the dinosaurs; and moreover, the dinosaurs preyed upon their own kind. Among today's living animals we see harmony; we see the many different species acting in such a way to promote ecological balance not benefiting one species to the detriment of others.

"In dinosaurs we see lack of harmony in the body plan, even though extreme sophistication can be noticed in some of them. Consider, for example, Tyrannosaurus rex's very short arms, the hands couldn't reach each other, there having probably no use for them; looking as if a mistake had been made and could not be fixed. The Pterosaur is another example of clumsiness; the animal could hardly stand and defend itself when on the ground. It's hard to believe that an animal like that could survive for a long time. In my opinion the dinosaurs were a part of a well-planned conspiracy against all life on this planet," commented Maryanne.

"You're right, Maryanne, there's something about those creatures that doesn't fit in the whole plan of creation. It seems that they came to destroy rather than to promote harmony," said Richard.

At this point Paula intervened again asking, "How could those creatures procreate, feed and grow?"

"They reproduced as birds do. They laid eggs, in some cases a hundred of them in the size of a soccer ball; and because of their resemblance to mammals; the carnivorous dinosaurs had a great appetite needing to eat a great amount of meat daily. According to some estimates, a carnivore had to eat yearly an amount of meat equivalent to fifty times its own weight; therefore a 1½ ton Tyrannosaurus would have to eat about 500 lb of meat a day.

"And according to David Norman, some scientists have discovered that because the dinosaurs had well-vascularized bones, which gave them more blood flow in the bone tissue, they could grow very fast. In some animals, it was observed that there was a reduction in the growing rate as they reached maturity, but in others the growing process seemed never to stop," Richard explained.

"But how about intelligence, were they intelligent or stupid animals?" Paula asked once more.

"Regarding the intelligence," said Richard, "the dinosaurs proved not to be as stupid as they were thought to be at first. If anything, they had brains comparable to those of modern reptiles, and a few of them were more active; it's possible that some of them had intelligence comparable to that of mammals and birds. And some scientists even believe that the theropods were developing relatively large brains (for dinosaurs); they could, if they had been given more time to follow this trend, have evolved into an intelligent species that could have dominated the planet. And I say that this was really the plan Lucifer had made for them, but the flood, coming at exactly the right time, put an end to it.

Chapter Six

The Gigantic Browsers

After a little pause, Richard resumed his explanation this time focusing his attention on the huge plant-eating dinosaurs, which were in his own conception nothing but powerful destroyers of the environment and creators of deserts.

"Dr. David Norman says that they belonged to the group of the *ornithischians*," Richard started, "ranging from the diminutive forms, the prosauropods, up to the gigantic sauropods. The prosauropods comprised the small and medium-sized dinosaurs, 13 to 20 feet long, capable of walking on all fours or on their hind legs alone. As time went on and they became heavier and bigger, they walked only on all fours standing on their hind legs only to reach for the higher foliage at the top of trees.

"The sauropods were the true giants of their time, and they included such notable creatures as *Diplodocus, Camarasaurus, Barosaurus* and *Apatosaurus*, more commonly known as *Brontosaurus*, all belonging to the same related group tending to have long, slender bodies, whip-like tails, long-shallow faces, and thin pencil-shaped teeth. Another scientist, Dr. David Gillette, says that all these giants lived in the western United States, likely roaming in herds and feeding constantly on conifers, cycads and ferns. They weighed on average twenty tons, with some of them, like the *Apatosaurus*, weighing considerably more. The giants existed in a great number, but there were also super giants, among which *Brachiosaurus*, that weighed twice as much as Apatosaurus – about seven to eleven times the weight of an elephant! Ultrasaurus, Supersaurus and also Seismosaurus were even larger than Diplodocus.

"The ornithischians had the hip bones very similar to that of birds, though scientists, confusingly, could find no family link in between them and birds. This doesn't seem strange to me. My opinion is that birds didn't evolve from dinosaurs, but in dinosaurs, those features which were so distinctive in birds, were methodically developed by the powerful mind of Lucifer using all the information on the DNA code that he could amass. We know that he could go much further than our most brilliant scientists can go today, due to his superior mind – a powerful genius that no human being can ever match.

"Dr. David Norman," Richard continued, "adds that among the plant-eaters there were some other medium-sized dinosaurs such as *Iguanodon* reaching the length of 30feet, and was particularly abundant in his time; *Hadrosaurs* in some cases grew to lengths of 43feet; they used to live in very large herds and were extremely efficient herbivores with special grinding teeth and muscular cheeks. Some scientists think that they were as numerous as the large herds of buffalo seen in the past on the North American plains, and of wildebeest that roamed on the African savannas.

"The *Ceratopians* like *Hadrosaurs* became phenomenally abundant as it is known by the enormous Ceratopians "graveyards" at some localities; and they too lived in large herds that roamed the plains of the Northern Hemisphere. These animals were well equipped with dense rows of teeth forming guillotine-like blades which could have sliced up the toughest of plants; they also had a parrot-like beak evoking their resemblance with birds."

"Considering the case of the much larger dinosaurs such as *Diplodocus*, *Apatosaurus*, *Brachiosaurus* that could eat approximately 450 pounds of foliage a day; and the Seismosaurus, the biggest of them all, 150feet long, which was equipped with a very efficient apparatus of teeth for chipping off leaves and small branches. They also had bird-like gizzards, inside which were gastroliths, apple-sized pebbles, to help the digestive process making it possible to them to assimilate nutrients from the toughest vegetation they could eat. These animals grew to a huge size, ate approximately a ton of foliage a day, and were apparently indestructible. They presumably could live up to 150 years, and in some cases laying up to 100 eggs each nesting season. It's no wonder that their population could grow to a very high numbers, since their huge size put them beyond the reach of any of the predators of their time; and even the Tyrannosaurus rex was unable to kill an adult, if it were not already injured.

"The herbivorous dinosaurs were planned in a great variety of sizes and shapes; had different ways of feeding on vegetation; some of them nested in colonies, tended their broods after hatching, feeding and protecting them. The larger sized dinosaurs fed upon the higher vegetation, while the medium and small sized fed on the medium and low vegetation. Everything seems to have been carefully planned in order to create as much desert as possible on the planet," Richard explained.

"How do they know how fast a dinosaur could grow and how long they could live?" asked Paula.

"After analyzing the bone structure of these dinosaurs and determining the large concentration of haversian canals through which the blood vessels irrigated and fed the bone tissue; and also the existence of a high amount of primary bone tissue in the dinosaur's legs, they were able to associate these characteristics with fast growing when the animal was still young. Some of them grew extremely quickly when they were young, reaching adult size in a surprisingly short number of years. Most of them slowed down or stopped the growing process after reaching maturity; but others seemed to have continued to grow actively throughout their lives, with no indication of any slowing down when they reached adulthood. As to their life span, they came to this conclusion after a careful examination of the bone structure observing the layers of the outer surface of the bone created by the aging and growing of the animal. Some dinosaurs could have a very short life span, while others could live even longer than a human being can live today," Richard answered.

"Dr. David Norman also says that some of the dinosaurs," Richard continued, "especially the medium and small-sized, could cope very well with drought due to their tough scaly skins, which protected them from dying out; they excreted little water, rather than losing a lot of liquid as urine; they produced a whitish, paste-like substance similar to bird droppings; and because they had some characteristics of a cold blooded animal, they didn't expend a lot of energy keeping themselves warm, and therefore could eat little and thrive where food was scarce and of poor quality."

"I wonder," commented Maryanne, "if those animals were able to thrive on earth as they were planned to do; they were tough creatures that could survive in the most unfavorable environment, maybe being capable of even supplanting humankind. Don't you think so, Richard?" she asked.

"Yes, I do!" he answered. "And there are some scientists," he continued, "who are specialized in the study of dinosaur copro-lites, or fossilized dung, and by studying these feces, they have been able to determine what kind of food those creatures ate. Of course, they haven't been able to associate the droppings to a specific kind of dinosaur, but at least they have been capable of guessing about the size and the fact that the animal was a carnivore or an herbivore. I wonder if they have found amidst the feces of those beasts at least a hint that humans and dinosaurs could have been contemporaries!" exclaimed Richard.

"What do you mean?" asked Paula, trying to understand Richard's reasoning.

"They have found some small pieces of broken bone mixed up in carnivorous dinosaur dung. Imagine if they suddenly found, in those droppings, something that could be identified as a human bone. Do you think they would ever publish this finding? I wouldn't be surprised if they didn't, since such a finding would strike a fatal blow to the theory of evolution. As far as I know, few are willing to sacrifice their most cherished beliefs and interests in order to buy the *pearl of great price*, the truth," he answered.

"You called the plant-eating dinosaurs 'destroyers of the environment and creators of deserts.' Why did you call them that?" asked Paula.

"It's well known that some regions are more likely to be transformed into desert than others; and that depending on how drastically the vegetation is damaged, specifically the big trees; and the quality of the soil. For instance, the Amazon Basin whose biggest part lies in Brazilian territory is densely covered by a tropical rain forest, the Amazon Forest. The soil, being not a very good, is unable to replace the native vegetation if it is cut, burned all at once or has large areas damaged. This is one of the reasons why the Brazilian government and the international community are deeply concerned about the preservation of that part of the rain forest. Without the abundant vegetation which grows there, there will be less rain in the area, causing the disappearing of most of the rivers that comprise the region's hydrography, consequently, endangering the rest of the world that depends upon the oxygen generated by the photosynthesis process (transformation of carbon dioxide [CO_2] breaking down its molecule and liberating oxygen [O_2]) that happens in the Amazon Jungle. Most of our oxygen comes from rainforests like this.

"So imagine what would happen if a herd of hundreds of hungry dinosaurs eating 400 pounds to a 1ton of foliage and branches a day, grazed in rain forests like the Amazon and others! Don't you think that it would take just a few decades until the entire region could be turned into a desert like the ones found in the African continent? And what would be the effect of a massive proliferation of deserts upon human life at the beginning of our civilization? (See appendix 2)

"The original plan in the beginning was for mankind to multiply and fill the earth subduing it; but this conspiracy involving the dinosaurs was something clearly designed to prevent human beings from becoming a dominant species on the planet, which was prepared and given to them as a gift from the creator. In the beginning the creator didn't have in mind that man should crowd themselves in the confines of the great cities. It was His purpose in the beginning that we should establish our homes in small communities, maybe little villages, in order to be more dispersed on the face of the globe in more contact with nature. The threat represented by those ferocious animals might have been one of the reasons why the antediluvian people decided that they should live in fortified, walled cities. Since then, throughout history, people have been largely concentrated in populous and overcrowded cities; and with the advent of the Industrial Revolution of 1825-1835 the process of human concentration in large cities was even more pronounced, giving origin to numerous large metropolitan areas such as New York city, in the USA, São Paulo, in Brazil, Mexico City, in Mexico, Tokyo, in Japan, and others; creating infrastructural problems such as inadequate mass transportation systems, precarious food supply and distribution, ineffective health care system, and high rates of unemployment; since the governments can not create jobs enough to absorb the newcomers that can be either the young people, who reach the working age, or the ones who come from other places seeking for better opportunities in life.

"Along with all these infrastructural problems we have to consider that the overpopulated cities are the hotbeds of all sorts of evil where homicides, prostitution, thievery, drug dealing, all sort of addictions and other evils, are perpetrated, and most of the time, go unpunished. In the cities of today, there's always turmoil, confusion, too much noise, and rushing about, impairing our ability to care for one another, and causing broken family relationships. In spite of that, most of us are strongly attracted to the life in big cities. The comfortableness, the social atmosphere, the intense consumption of goods and the easy lifestyle afforded by a large variety of commodities; all of these exert a powerful influence on families, while the true evil of life in overcrowded cities is ignored.

Adding to all that, the moral and social problems that can generate so much evil that men and women of sound and righteous mind are forced to conclude that we live in a sick society whose maladies can not be cured by the adoption of correct and effective methods of administration alone, but also by the education of humankind in the principles of righteous living as delineated in the Word of God since the beginning of our civilization," Richard concluded his long and meaningful answer.

"Not only that," commented Paula, "the physical surroundings of the great cities are often a peril to health; and the constant exposure to disease; the prevalence of foul air, impure water, and impure food, in addition to the crowded, dark and unhealthy dwellings are some of the evils of life in large cities."

"Well," commented Maryanne, "this guy, Lucifer, is not playing around. In my opinion he's firmly determined to destroy as many human beings as he can; and he'll never make it easy for any of us to attain the purpose God has for each one of us. If we are to be on the Lord's side, we have to do it with intelligence, determination and the support of His grace, otherwise we'll be swept away by the enemy's power. Don't you think so, Richard?" she asked.

"Yes I do think so. All those who will be standing in the end, will be the ones who washed their robes in the blood of the Lamb, entered in a covenant relationship with Him and were enabled to overcome the dragon, the beast and its image," said Richard quoting Revelation 13:11-18.

Chapter Seven
The Strategy

Paula was so interested in that conversation about dinosaurs that for a while she forgot she was hungry completely ignoring the appetizing smell of the food.

Mrs. Campbell invited everybody to sit around the table where the food was set out for luncheon, as Mr. George Campbell was invited to say a word of prayer in gratitude for the food and the special ingathering of the family. Richard, who had already finished roasting the barbecue, was asked to bring it over to the table; and as he left the grill, he said to the two girls, "Don't go anywhere, I'll be right back! I still have something very special to tell you about the dinosaurs," he added.

"Don't worry, we're not going anywhere!" answered Paula with a smile.

Richard didn't take too long to come back to the girls; and when he returned, he brought them two plates filled with salad, white rice, some barbecue, and brought each a cup of soft drink filled to the top. The two girls were so delighted that they couldn't hide their emotion and kissed him on the cheek.

"Thanks, cousin!" said Paula as Maryanne was laughing loudly.

"Now you can continue the lecture, my dear brother," said Maryanne glad to be his sister and proud of his intelligence.

"Ok," said Richard, "let me resume my explanation. In the beginning mankind was meant to procreate, fill the earth and subdue it. The animals were placed under Adam's care and authority; he gave them names, and all of them were subjected to him. He was supposed to be the representative of mankind before the other worlds; a privilege that he lost when he fell into sin. Afterwards, what we see is an attempt to subvert the plan God had made for man; a plan that would elevate our species to immortality; in the beginning God had given man conditional immortality depending upon his obedience to His commandments."

"How, do you think, the initial plan for mankind was attacked?" Paula asked.

"The initial command that man should fill the earth and subdue was first attacked when Lucifer developed the dinosaurs, spreading them all over the world. And if we look at the map that tracks dinosaur bone findings, we see them distributed all over the globe. Dr. David Norman, in his book *Dinosaur!* says that dinosaur remains have now been found on every continent on the globe. Dinosaur fossils have even been discovered in the frozen wastes of Antarctica by teams from Britain and Argentina. Western Europe and North America, the traditional areas for dinosaur fossils, are still producing important new finds. However, exciting discoveries are being made particularly in Mongolia, China, South America, Africa and Australia.

"It's interesting to notice that when you start paying attention to where they are distributed and the specimens they've found, you can see that there can be two explanations for these dinosaur fossil distribution; either they were developed, and then placed in strategic places to induce people to think of them as evolutionary steps of each kind of dinosaur, or their apparent succession in the evolutionary line is just an evidence that their developer, Lucifer, didn't have them all defined in his mind, not knowing exactly how to adjust the DNA code in order to command the protein-maker machine, in the dinosaur cell, to produce the specific proteins that would result in the development of the planned anatomical and organic features in those creatures. It seems that after many unsuccessful attempts, he got some of them in the way he wanted. Consider, for instance, Postosuchus, an attempt that failed; even though it had all the potential, it wasn't yet a dinosaur the way they define it today; Coelophysis, considered one of the first dinosaurs; Herrerasaurus, considered by some as a proto-dinosaur because it doesn't fit among the species known at present; though, according to specialists, there were several hundred species already listed, varying greatly in size and shape," Richard concluded. (See appendix 3)

"Are you saying that Lucifer tried to develop them little by little, doing his experiments until he could get the results he expected?" Paula asked.

"Yes, that's what I mean. In other words, when he started, he didn't know exactly what he would obtain in the end, so he made experiments until the desired results were obtained," answered Richard.

Maryanne, who was paying close attention, suddenly, moved by an intense curiosity, asked, "What do you think was his plan in developing and placing them strategically on the planet?"

"I think he was trying to limit the area of human habitation, since he placed the dinosaurs all around, except for the area where civilization first appeared, Mesopotamia and the Middle East. If you look at the map of fossil distribution around the globe, you won't see any dinosaur bones being found in that area; and with the development of the carnivorous dinosaurs, he might have meant to destroy whoever would venture out of that area. I suppose that those who lived inside the *restricted area*, Mesopotamia and Middle East, may have never seen a dinosaur like Tyrannosaurus rex, or one of the gigantic herbivores such as Seismosaurus or Brontosaurus. Only those courageous men who would dare to cross the border line and were lucky enough to come back, were able to bring the story about the creatures they had seen in those distant lands.

"Earth in its primeval condition was covered by luxuriant vegetation composed of a great variety of flowers, majestic trees, palm trees and shrubs, all this designed to support human life; but with the development of the huge herbivores, paradise was doomed and surely would soon be transformed into a scalding desert. Considering that some of those animals could live for about 150 years and eat up to a ton of foliage a day, imagine what would happen to the environment whenever a herd of those hungry creatures browsed in a certain area. It would certainly be turned into a desert. Scientists today have strong evidence which indicates that those gigantic browsers traveled long distances looking for the grazing areas and nesting grounds; and as they moved on from place to place, all that was left was only a denude and arid land. I believe that if Lucifer had been allowed to develop his plan to its full extent, the entire planet would soon be reduced to a smoldering desert, culminating, most likely, in the extinction of the human race," Richard paused a little bit.

"But if they were destroying the environment wherever they went, wouldn't that also cause their own demise, since they were destroying all the vegetation, their only source of nourishment?" Paula asked.

"The dinosaurs were tough animals, and some of them could withstand long periods of drought and hunger, traveling very long distances in search for food and water; and only a calamity with the proportions of a worldwide flood could have killed them once and for all," Richard answered.

"Another aspect of Lucifer's strategy," he continued, "is related to the social and moral problems generated by high concentrations of people in the confines of walled cities. The Bible account tells us about a man named Nimrod who was dubbed '*a mighty hunter before the Lord*,' who was also the founder of four great cities: Babel, Erech, Accad and Calneh in the land of Shinar. He began one of the greatest empires on earth, the Babylonian Empire. Nimrod lived after the flood; therefore, after the dinosaurs had been killed; but the idea of living in big cities existed even before the deluge, and the story of the antediluvian huge animals might have caused a strong impression upon people's minds, leading them to decide not to take any chances, and to build the Tower of Babel, whose top would touch the skies – so they could build themselves a name, and avoid being scattered all over the world – and if another flood should occur, they would be safe climbing to the top of the tower.

"Now that Lucifer had them living in walled and overcrowded cities, it would be easier for him to work on man's character trying to deprave it the most; leveling it, if possible, with the brutishness of beasts. As an ancient example of corrupt cities we have Sodom and Gomorrah, the cities of the plain, that were populated by rich people who didn't have to work very hard for their subsistence, since the land was rich, the soil fertile, and all of them had more than enough to provide for themselves. The great prosperity and ease of these two cities, were the main reason for their downfall; because when the mind is free from the worries and hardships of a difficult life, people almost always start thinking about the sensory pleasures of life such as partying, drinking, lasciviousness, sex, drug abuse, gambling, and other excesses. That, supposedly, was the case of Sodom and Gomorrah. Ellen G. White in her book *Patriarchs and Prophets*, comments on the life in the cities of Sodom and Gomorrah: 'In Sodom there was mirth and revelry, feasting and drunkenness. The vilest and most brutal passions were unrestrained. The people openly defied God and His law and delighted in deeds of violence. Though they had before them the example of the antediluvian world, and knew how the wrath of God had been manifested in their destruction, yet they followed the same course of wickedness.' Afterwards she compares the cities of today with the cursed ones of the plain: 'The cities of today are fast becoming like Sodom and Gomorrah. Holidays are numerous; the whirl of excitement and pleasure attracts thousands from the sober duties of life. The exciting sports, theater going, horse racing, gambling, liquor drinking and reveling, stimulates every passion to activity,' *ibid*. And she adds another comment on life in overpopulated cities: "Every day [life in a big metropolis] brings fresh revelation of political strife, bribery and fraud; every day brings its heart sickening record of violence and lawlessness, of indifference to human suffering; of brutal, fiendish destruction of human life. Every day testifies to the increase of insanity, murder and suicide,' *ibid*.

"The greatest depravity is frequently found among the people living in the most prosperous lands, since iniquity seems to go very well with prosperity. No matter if it's a nation, a city or a person; the fact is that where the gifts of God are abundantly lavished, there men forget Him first. And most of the time, self-satisfaction brings along with it one of the moral dangers that threatens all of those endowed with material prosperity, leading them to such a degree of involvement with the things of the present world that they feel no need of God.

"Sodom and Gomorrah became proverbial for their iniquity, even though they had been greatly blessed with the advantages of a more advanced civilization; and today the inhabitants of the two cities are not remembered for their high degree of education or any achievement that has contributed for the progress of humankind, but for their wickedness and depravity that gave rise to perverted sexual practices, sodomy, that characterized the inhabitants of the two cities of the plain.

"That is one aspect of the strategy used by Lucifer in order to secure the deterioration and destruction of the human family. He knows that immorality and addiction act strongly upon the mind, weakening the will and sapping the structure of the character. He was successful in the past and he still uses the same method, and that tells us about the kind of enemy we have to face in our every day life. He is subtle and insidious, acting so stealthily that no one can detect his actions; being his sphere of influence everywhere, encompassing the whole world. He can influence an individual's life, whole families, and the most elevated people in strategic positions in the government, having them to approve laws that will promote the growth of his kingdom, making life more difficult for those who want to obey God's commandments, walking in the way of justice.

"His subtlety is so refined that even the professed Christians don't quite understand his way of deceiving. He loves when people, in their ignorance, depict him as a horny creature, half man and half animal with a long and horrible tail; since behind this ridiculous imagery, he can hide his angelic appearance, and go unnoticed as he works out his evil plans to overthrow God's throne and destroy us. That is what he really is, an angel; and as such he is greatly superior to human beings. Being known as the master of deceit, he was able to deceive one third of the heavenly host and woo their sympathy towards his cause. The dinosaurs and the strategy involving these huge beasts are only a tiny part of his plan for the destruction of human race; and right now, at this very moment, he has his powerful agencies carrying out his plans of deception and destruction all over the world. He is a kind of 'hard worker' whose feet are fast to do evil," Richard explained, pausing for breath afterwards.

Maryanne asked, "Do we have any chances to withstand his attacks, since he is so powerful?"

"Yes," answered Richard, "he is powerful, but God is infinitely more so. Remember that he's a being, created by God, and as such he has limits to what he can do. The Creator established his limits and he can't go beyond them; his life is derived from God and his power as well; and he has only been allowed to develop, before the expectant universe, all the evil generated by his rebellious mind; and, then, he will be revealed before all the intelligent beings, who inhabit the universe, as he really is, a deceiver, a murderer and the one who's been plotting against human beings since the beginning. The flood was the means that God used to put an end to the plot involving the dinosaurs, but since then the enemy has never stopped his work, and with a strong determination, he still keeps trying to destroy as many human beings as he can; and our only protection is to be under the protective hands of the Almighty."

Chapter Eight

A Profitable Business

Paul Silvers was beginning to get impatient. He had been waiting for his classmate, Richard, for about an hour, and he began to worry. "Will Richard come?" he wondered. That was an unusual delay because, as far as he could remember, his buddy had never been late for an appointment; and that one was important, since they were supposed to meet at the library and spend some time studying for the next-day-calculus test. In Paul's mind it was very unlikely that Richard would miss such an appointment for no significant reason. "What happened to him?" Paul asked himself. "Did he miss the bus? Did he get stuck in a traffic jam? Did he oversleep this morning?" Finally, realizing that all these possibilities could have occurred, he decided not to worry and, instead, wait patiently for his friend. After an hour had passed, Richard Campbell finally got there; breathing a little hard in his hurry.

"Hello!" he said.

"Howdy!" replied Paul.

"What happened to you? Why are you so late?" asked Paul, showing a little bit of irritation. "We have that calculus test tomorrow; it'll be a hard one, maybe the hardest we've ever taken. Now, we'll need all the time available to study!"

"Yes," agreed Richard, apologizing and trying to explain his lateness. Paul told him it wasn't important, so they decided to begin studying for the test. They spent a long time reviewing the sample problems; some of them being very difficult to solve, which didn't represent a serious problem to the two young men, since they had plenty of notes taken in class. After a long while, they took a break to refresh their minds, and decided to take a stroll around the campus ending up in the cafeteria, where they sat down for a drink and a little chat.

"How are you doing with your research on dinosaurs, Richard? You've been reading about them for so long that you should be an expert on the subject by now, shouldn't you?" asked Paul to his friend playfully.

"Well, I'm not as knowledgeable as you might think. My knowledge on the subject still has many blank spaces to fill in. In spite of all that, I can say I pretty much understand it in a level that gives me condition to have my own explanation on what happened to those huge and dangerous creatures," said Richard with a serious tone in his voice.

Paul was noticeably touched by Richard's words; not only the words, but by Richard's serious tone of voice; specially taking into consideration how Paul admired the younger man's intelligence. They had been friends for a long time, even before they started taking the Mechanical Engineering course at the same university. Paul used to tell to his family and friends that Richard was one of the most intelligent persons he'd ever met. He liked Richard's sound reasoning and clear line of thought. Anything explained by Richard would be easily understood, even by those of average intelligence, such was the young man's sharpness when it came to explain his ideas and make his points on any complicated matter.

"And what is your explanation of what happened to the dinosaurs?" asked Paul teasing him.

"Well, before I start explaining my theory, I would like to mention how this dinosaur thing has become a very profitable business today; and one doesn't need to be extremely brilliant to notice the significant amount of material about those huge creatures presently in the market. I can effortlessly remember the title of three or four recently released documentaries that are still selling a great deal; but let's not consider only the videotapes. How about the movies, cartoons, books, magazines and the great variety of toys that are introduced into the market almost on a daily basis? I wonder how many people are choosing university and college courses related to the field of paleontology motivated by all this specialized talking about our *fossilized friends*. And if that is true, the merchandise retailers are not the only ones making profit on this dinosaur market, but also researchers, teachers, educational institutions, and all the other professionals who work in obscurity doing the job that demands less technical qualifications, assisting those who step on the stage and actually run the *show*. All I've said so far is only a preamble of the main point I want to make: It's very hard, maybe impossible to the system to accept my theory; and the reason for that is that it would make things too simple, and therefore, uninteresting. It wouldn't sell merchandise and ideas as much as the established theory does now; and not everybody is willing to sacrifice a generous income from the evolutionary dinosaur theory, to buy the *pearl of great price*, the truth, because it has economical, moral and spiritual implications. Besides all that, I must say, the main issue in discussion is not 'who is saying' or 'what is being said'; but who is behind this entire story since it started – Lucifer, the cherub, who once was an angel of light, the morning star, and the one whose face once reflected the glory of God. He is the one who's been plotting against mankind trying to snatch us from God's protection and wipe us out. But there's still an apparent contradiction that I want to clarify. While believing that, maybe, just a few people will share my convictions, I still talk about it, hoping that somewhere, somebody will think it over and get the correct idea about Lucifer, who's been sponsoring all this erroneous interpretation of the story, holding people in ignorance of his criminal actions ever since the beginning of history. If that happens, I'll have achieved my goal and then I'll be satisfied," Richard concluded.

When the younger pal had finished the first part of his explanation, Paul clapped his hands in a gesture of approval and exclaimed, "Bravo! Bravo! Bravo! Now, Richard, listen to me," he spoke with a serious intonation in his voice, "I know your religious convictions, and I particularly admire you; I partially agree with your theory about dinosaurs. Ok? But my question still remains unanswered. What happened to the dinosaurs?"

"Well, let's start by saying that I partially agree with what the paleontologists are saying about those creatures. They have made so many important discoveries about their habits; their way of hanging around in groups, some of them forming large colonies; their way of reproduction by laying eggs; the way they took care of their offspring; their habit of traveling in herds, with their young on the inside of the group for protection. But one of the most significant facts about dinosaurs that the scientists of our age have discovered is related to their having the characteristics of cold blooded and warm blooded animals, so perfectly combined, giving them the best features of mammals and reptiles whenever it was convenient to their survival and the purpose for which they were *developed* which was to be used as killing machines to wipe out human beings," explained Richard to his attentive friend.

"Hey! Wait a moment! I know that you are a religious person, and I would expect to hear you say that the dinosaurs were created by God instead of having '*evolved or being developed*' as you just said," commented Paul trying to get the point of Richard's words.

"Don't be so quick to draw conclusions out of my words, Paul!" exclaimed Richard, "I didn't say that they were not created, but that they were *developed*; and by that, I mean that since only God can create life, I'm suggesting that Lucifer designed them and, by genetic manipulation, *developed* them, giving them the anatomical and organic features he wanted, to make them into killing machines with the definite purpose of annihilating humankind," said Richard, reinforcing his point to make it clearer.

"How could that be?" asked Paul, full of curiosity.

"Today it's a fact well-known, that, by genetic manipulation, clones can be produced – being not certain, however, if they have already successfully tried it on human beings, even though there have been some scientists claiming to have, recently, produced a human clone – and that the things which are possible to be done by humans with their limited minds, were much easier to be achieved by a powerful mind as Lucifer's, even in the past. He knows genetics more than any of our best scientists can ever know; and most of the ideas and concepts we strive to grasp, he knows them by heart. I wouldn't be surprised in knowing that he started making his experiments with *Postosuchus*, an early stage in the development of dinosaurs that failed, considering its posture and the way it moved around; *Coelophysis*, considered the first land dinosaur; and *Deinonychus*, a little bit bigger than a *Velociraptor*; working his way up to the most perfected killing machine that we've ever heard about, *Tyrannosaurus rex*, a carnivorous dinosaur whose length could reach up to 50 feet, from the head to the tip of the tail, and a height of approximately 20 feet; *Diplodocus*, *Brachiosaurus*, *Seismosaurus* and other super giants that could eat up to a ton of foliage in just one day.

"Most scientists today – with maybe a few exceptions, I think – believe that birds evolved from dinosaurs, based on the similarity of the hip bones between certain kind of dinosaurs and birds, which is not true, because we know how birds came into existence, while the dinosaurs still remain a mystery, with nobody knowing for sure how they began, since there's no biblical reference about their creation. But one thing we can infer: Birds, besides humans and maybe a few other species on this planet can stand on two legs and move around with considerable equilibrium and speed; what gives them great advantage when compared to other animals. Knowing that, it's not difficult to understand that whoever sets out to the task of designing a new species, having a certain purpose in mind, would at least try to develop in this new animal some of the best features of the most successful species on the planet. That was the case with birds. They had some characteristics that a powerful killer like *Tyrannosaurus rex* or other predators should have, such as the ability to stand on two legs, good balance, and speed.

"So, in my understanding, the interpretation goes all the way around. Birds didn't evolve from dinosaurs; but on dinosaurs, those features which were so distinctive in birds, were methodically developed by Lucifer's powerful intelligence, which is vastly superior to ours, and whose capability we can't accurately measure, since our limited minds can only glimpse his accomplishments and power." (See appendix 4)

"Finally, when he had developed all the dinosaur species which were most dangerous to the environment and human beings and was ready to start accomplishing his evil intents, the Flood came and destroyed them all, humans and dinosaurs alike, leaving only one family – Noah's – to repopulate the earth; meaning that it is not the physically superior who will be able to survive in the end, but the faithful who fulfills his part in the plan of the Creator."

Chapter Nine

The Flood

Some weeks after Aunt Millie and her daughter Paula had visited the Campbells on that sunny Sunday morning; the family had another opportunity to get together again, this time to celebrate Richard's birthday. There was excitement everywhere and among those present were Paul Silvers, his girlfriend Martha, and Paul's parents, Robert and Amanda Silvers. The small house was overcrowded that evening when Richard was celebrating his 22nd birthday. Now that he had gained more maturity, his reasoning was even sharper than it used to be before in his years of adolescence, being now more capable of making his points with more clarity and conciseness than ever before.

George Campbell, Richard's father, seemed to be having the best day of his life; and Mrs. Campbell, though extremely happy about the celebration, showed no excitement, or nothing that could suggest the slightest lack of control of the situation. She talked and walked among the guests with scintillating eyes, a smile on the face, revealing her enjoyment of her son's birthday celebration.

Maryanne, who seemed more excited than everybody else, walked to the middle of the living room, where some of the furniture had been moved aside so people could have more room to stand and talk; and then said, "Folks, I would like to have your attention for a moment. Today, as you know, is my brother's birthday, and I'm very happy because he's so dear to me; but my father has something to say. Please give him your undivided attention."

George Campbell was a man of few words, and as such he had the habit of using them in the most simple and effective way he could. He said, "Birthday celebration is one of the best ideas that man could ever have had. It tells us that time is passing; bringing experience along with the aging of our bodies; and it also tells us that one day we'll pass away, as the grass that comes forth, flourishes, and after a short while dries up and dies. Such is human life. It has beauty in its beginning, and as the years go by, it loses strength and goes its way. Aging is an important aspect of life; we can't help getting older, but we surely can take advantage of the knowledge and experience that come along with it. Richard, now, is just twenty-two, and it seems that he doesn't have any reason to be concerned about getting old, because it's going to take a while until he gets there; but I say if he's concerned about that today, he certainly will be better equipped to deal with it when the time comes."

"What do you mean by 'being better equipped', Mr. Campbell?" Paul asked.

"I mean that if you live your present life the right way, using your intelligence and seeking God's guidance as you make your decisions, you can't go wrong, and in the end you'll still find happiness in life," replied Mr. Campbell. "Let us bow our heads and have a word of prayer in gratitude for Richard's life," he added.

They prayed, and afterwards the guests resumed their conversations talking about a wide variety of subjects. Then Richard, attracted by the interesting conversation of the group made up by Maryanne, Paula, Paul Silvers and his girl friend Martha, finally decided to join them.

Paul Silvers, as he saw Richard approaching, took the initiative and said, "The last time I and Richard talked, it was in the library, and we had a very interesting conversation. Richard told me about his theory on dinosaurs – how they came about and what caused their extinction. We didn't have much time to talk because we had to prepare for the calculus test next day; but now, if you don't mind, folks, maybe our young lecturer could finish that explanation and give us more information on the dinosaurs' demise." The girls agreed and Richard started.

"When Lucifer had developed the most dangerous and lethal carnivores, such as *Tyrannosaurus rex*, *Deinonychus* and *Velociraptors*, and the super giant herbivores, such as *Brachiosaurus*, *Diplodocus*, *Brontosaurus*, *Seismosaurus* and others; the flood came and all of them and the doomed generation of corrupt antediluvians were destroyed at once," he explained.

"By that time," he continued, "the idea of a flood that would destroy all living things upon the earth was something *crazy and hard to believe* for those who were not led by the Spirit of God. In the beginning, when our planet was still in its infancy even after the fall of Adam and Eve into sin, said Ellen G. White in *The Great Controversy*: 'nature still kept much of its beauty, with majestic trees being vastly larger surpassing in beauty and perfect proportions anything that mortals can now look upon. The wood of these trees was of fine grain and, in this respect more like a stone growing upon the highest elevations, rising to lofty heights, their branches spreading to a great distance on every side, while the plains were covered with verdure and appeared like a vast garden of flowers. Some of the hills were covered with trees of beauty and vines climbing the stately trees were loaded with grapes, while beautiful flowers filled the air with their fragrance.' The men and women living at the time were of very great stature possessing wonderful strength, being their life expectancy much longer than the one we have today. Man could live hundreds of years and still be strong in his old age. And when Noah came up with the idea that the world would be destroyed by a flood, the great thinkers of the time scorned and dubbed him a *crazy old man*. Ten generations had passed since Adam; and nobody had ever seen rain fall upon the earth; so they reasoned that rain was something impossible to happen and the idea of a flood was totally absurd. Absurd or not, the deluge came at the appointed time, and those who were unprepared were destroyed by it.

"Contrarily to what many people think, the flood wasn't only local, but covered the entire face of the earth. The Hebrew account of the flood, as presented in the book of Genesis, reveals the tremendous extent and intensity of the cataclysm by a graphic series of verbs and adverbs: the waters 'increased'; 'prevailed' and 'increased greatly'; prevailed exceedingly and even 'prevailed' fifteen cubits, or twenty-six feet above the *highest* mountains. The description is simple, majestic and vivid; an immensurable volume of water covered the whole earth (*SDABC*-page 257).

"The global scale of this catastrophe is also attested by the flood legends preserved among people of nearly every race on the face of the globe; being the most complete of these accounts that of ancient Babylonians, who settled in proximity to the place where the ark came to rest after the flood, from whence the human race again began to spread abroad (ibid.). He continued, "On December 3, 1872, an Assyriologist named George Smith presented in a paper to the Society of Biblical Archaeology the translation of Tablet XI of the Gilgamesh epic, written in Akkadian, an ancient Semitic language, older than Hebrew, in which he talks about a flood story that had remarkable parallel to the biblical deluge. The Epic of Gilgamesh, the Sumerian king who reigned in the fortified city of Erech, founded by Nimrod the phenomenal builder and *mighty hunter before the Lord*, bears many similarities to that of Genesis, and yet differs from it in such a way as to prove the narrative of Genesis as something that could only have come from the inspiration of God.

"In the Genesis account, the wickedness of man is presented as the reason for the flood which destroyed all living things, including beasts and man; but parallel to that Lucifer had been tampering with animals and human beings, obtaining by amalgamation the development of weird aberrations of men and beasts; and I, personally, have a theory about this. The fossils known as "Homo erectus", "Neanderthal man", "Australopithecus" from which they say, the modern man, Homo sapiens has evolved, were all results of Lucifer's tampering with humankind that he planted in strategic places afterwards, as he did with the dinosaurs, in order to give the impression that they were steps in the evolutionary line of human beings. Regarding the animals that were placed in the ark, it is said that of every kind of animal that *God had created* there were two pairs of the unclean and seven pairs of the clean; so the flood didn't come only to destroy an entire generation of perverted men, but also to wipe out the huge and ferocious animals that *were not created* by God, such as the dinosaurs and other weird creatures that Lucifer had developed by amalgamation. The dinosaurs were the reason for special concern in this case; since the fallen archangel had successfully developed those species that could destroy human life and its support – flora, fauna and environment – creating a serious difficulty to human existence on the planet. The plan, I believe, wasn't only to kill human beings; but by confining them in walled cities, corrupting them morally and physically, to destroy the family structure and finally the individual," Richard concluded. (See appendix 5)

Richard paused and, then, continued, "When the rain started to fall from the clouds above," says Ellen G.White, "this was something that the men had never witnessed and their hearts began to faint with fear. The beasts were roving about in wildest terror and their discordant voices seemed to moan out their own destiny and the fate of man. The storm increased in violence until water seemed to come from heaven like mighty cataracts. The boundaries of rivers broke away and waters rushed to the valleys. The foundations of the great deep also were broken up; and jets of water would burst up from the earth with indescribable force, throwing massive rocks hundreds of feet into the air and then they would bring themselves deep into the earth.

"The people first beheld the destruction of the works of their hands. Their splendid buildings, their beautifully arranged gardens and groves where they had placed their idols, were destroyed by lightning everywhere.

"The violence of the storm increased, and there was mingled with the warring of the elements, the wailings of the people who had despised the authority of God. Trees, buildings, rocks and earth, were hurled in every direction. The terror of man and beast was beyond description; and Satan (Lucifer) himself, who was compelled to be amid the warring elements, feared for his own existence. He had delighted to control so powerful a race, wished them to live to practice their abominations, and increase their rebellion against the God of heaven. He uttered imprecations against God, charging Him with injustice and cruelty.

"Some of the men clung to the ark until borne away with the furious surging of the waters, or their hold was broken off the rocks and trees; and they were hurled in every direction. The animals exposed to the tempest rushed toward man choosing the society of human beings as though expecting help from them. Some of the people would bind their children and themselves upon powerful beasts, knowing that they would be tenacious for life and would climb the highest points to escape the rising water. The storm does not abate its fury – the waters increase faster than at first. Some fasten themselves to lofty trees upon the highest points of land, but these trees are torn up by the roots and carried with violence through the air, and appear as though angrily hurled with stones and earth into the foaming waters, which nearly reached the highest points of land. The loftiest heights are at length reached, and man and beast alike perish by the waters of the flood.

"Before the flood there were immense forests. The trees were many times larger than any trees we now see. They were of great durability and they would know nothing of decay for hundreds of years. At the time of the flood these forests were torn up or broken down and buried in the earth. In some places large quantities of these immense trees were thrown together and covered with stones and earth by the commotions of the flood. They have since petrified and become coal which accounts for the large coal beds which are now found."

"Everywhere over the surface of the earth we find fossilized remains of plants and animals obviously deposited by water, being the universal distribution of these remains and the depth of their burial an unmistakable testimony to both the worldwide extent and the terrific violence of Noah's deluge.

"The whole surface of the earth was changed at the flood... The beautiful trees and shrubbery bearing flowers were destroyed...

Soon after the flood, trees and plants seemed to spring out of the very rocks. In God's providence seeds were scattered and driven into the crevices of the rocks, and there securely hid for the future use of man."

"Presently," Richard adds his personal observation, "in the United States and Canada there are two dinosaur sites that are of note due to some things found in these places that suggest the worldwide occurrence of the flood. One of them, the Dinosaur National Monument, located in Utah, has an exposed wall of sandstone that has been carefully excavated to reveal about 1,500 dinosaur bones. The wall shows a partial skeleton of a young Camarasaurus, along with isolated bones of several other dinosaurs such as Apatosaurus, Diplodocus, Allosaurus and Dryosaurus. To date, ten different genera of dinosaurs have been found at the site characterizing it as the one that has the largest variety of dinosaur's species anywhere in the world. The richness of the fossil finds suggests that the dinosaurs were probably swept away and drowned by a large amount of water. The other site under consideration, the Provincial Park in Alberta Canada, is just as intriguing, having yielded over 300 skeletons of about 35 separate species of dinosaurs. This bone bed discovered in 1970 has produced about 60 bones per square meter; being found small bones from specimens measuring between 3 to 4 feet long, and long ones from specimens measuring about 18 ft long; with 85% to 95% of the bones belonging to one single species, the Centrosaurus. But even more interesting is the fact that the bones are fractured, indicating the possibility that the beasts could have been hit by a strong blow just before the moment of death as indicated by the fractures which spiral through the bones, instead of fracturing the bones straight across; making scientists believe that they were broken when the animal was still alive. In this case, too, the occurrence of a flood with the magnitude and violence of that of Noah's time, as described by the inspiration of Ellen G. White, seems to be confirmed."

Chapter Ten
Final Words

Richard paused a little bit in his explanation; as if he wanted to give somebody a chance to ask questions or make any comments on what he had just finished explaining. Then Maryanne spoke up and said, "Richard, I was just wondering how this plot – the dinosaur thing, you know – is related to the great controversy between good and evil. Is the plot a part of this great controversy?"

"Well," Richard said, "I personally believe that it is; even though I also believe that the plot is just a tiny part of it."

At this point Paula, who was attentively listening to all that Richard had said so far, interrupted with a question: "Richard, what is this great controversy? Would you mind explaining to me what it is?" she added.

"Yes, of course!" he answered. "In the beginning," he continued, "at a point in time before the creation of other beings such as the angels and the inhabitants of the unfallen worlds; there was only God, who being himself the greatest expression of love and motivated by this love, decided to create intelligent life so He could share with his creatures all the blessings and happiness of existence. But the creator, as we know, besides being omnipotent and omnipresent, is also omniscient, which means that He knows the end since the beginning, there being no secrets before His eyes. He foresaw the coming of evil, since freedom of choice would be a paramount in the new order He was just about to establish. His creatures, as intelligent beings, would have freedom to choose between good and evil, right and wrong. The fact that he could see that somebody like Lucifer at some point would choose the way of injustice, didn't cause Him to change His mind about creating intelligent beings to populate the universe.

"That was the time, in a distant point of eternity, before the creation of all intelligent life, when the plan of salvation was established in order to face a problem that would arise in the future. The plan of redemption caused a change to occur in the very way God, previously, manifested Himself; because in order to carry out the plan, now He manifested Himself in the persons of the Father, the Son and the Holy Spirit – the Trinity – making it possible for Him, in the person of Jesus Christ, to take upon Himself human nature and die on behalf of the fallen human race. The most amazing thing in all this being the fact that He didn't give up the idea of creating us; and instead, He loved us even before we existed and gave us the chance to come into existence and enjoy all the blessings that come with the gift of life. Referring to Jesus Christ's incarnation, the Bible, in the book of Isaiah 9:6, states: "For unto us a child is born, unto us a Son is given; and the government shall be upon his shoulder; and his name shall be called Wonderful, Counselor, The Mighty God, The everlasting Father, The Prince of Peace. Yes, Jesus Christ is the Son of God and also the Son of Man being forever a representative of our race in the heavenly courts where he will be seating on the throne of the universe. This is our privilege and blessing; a human being, in the person of Jesus Christ, will rule over the entire creation of God.

"Ellen G. White, in her book *The Great Controversy*, gives some information on how perfect Lucifer was before sin was found in him: 'Satan – Lucifer – in heaven, before his rebellion, was a high and exalted angel, next in honor to God's dear Son. His countenance, like those of the other angels, was mild and expressive of happiness. His forehead was high and broad, showing a powerful intellect. His form was perfect; his bearing noble and majestic. A special light beamed in his countenance, and shone around him brighter and more beautiful than around the other angels...'

"Even though Lucifer, as the most elevated among the angels, had a special and privileged position in heaven, he was still envious. Ellen G. White continues: 'Yet Jesus, God's dear Son, had the preeminence over the entire angelic host. He was one with the Father before the angels were created. Satan was envious of Christ, and gradually assumed command which devolved on Christ alone' (*ibid*.).

"In face of these feelings of discord, which started growing in Lucifer's heart,' Ellen White explains, 'The Great Creator assembled the heavenly host, that He might in the presence of all the angels confer special honor upon His Son. The Son was seated on the throne with the Father, and the heavenly throng of holy angels was gathered around them. The Father then made known that it was ordained by him that Christ, the Son, should equal with himself; so that wherever was the presence of his Son it was as his own presence. The word of the Son was to be obeyed as readily as the word of the Father. His Son he had invested with authority to command the heavenly host.'

"Not only was the Son placed in a high position of command of the entire angelical host including Lucifer who was told to reverence his presence and bow down before him in adoration; but He was also entrusted with the responsibility of working with the Father in the creation of the earth and of every living thing that should inhabit the planet. 'His Son would carry out His will and His purposes, but would do nothing of Himself alone. The will of the Father would be fulfilled in him,'" (*GC*-18; *See also John 1:1-3; Genesis 1:26, 27; Colossians 1:16*)

"The Father consulted with Jesus in regard to at once carrying out their purpose to make man to inhabit the earth. He would place man upon probation to test his loyalty, before he could be rendered eternally secure... He did not see fit to place them beyond the power of disobedience." (*GC* 23)

"Lucifer thought himself as being a favorite in heaven among the angels and the consciousness of that didn't generate in his heart a feeling of gratitude to the creator for the blessings and high privileges he enjoyed in the heavenly courts, instead it created in his heart the desire to be in the same height as God." (GC 18; Isaiah 14:14)

"If one asks for a reason why somebody so highly gifted as Lucifer would choose to rebel and seek to destroy his own creator, it would never be presented. Lucifer had no reason to justify or even explain his actions, since the attempt to explain sin is sin itself. Every thing was perfect, there having no mistake in God's government, except that pride and the desire of self glorification were found in Lucifer's heart who used all the gifts and endowments imparted to him to advocate his rebellious cause and, by deceit, win one third of the heavenly host to his side. Ellen G. White says, 'Good angels wept to hear the words of Satan and his exulting boasts. God declared that the rebellious should remain in heaven no longer. Their high and happy state had been held upon condition of obedience to the law which God had given to govern the high order of intelligences. But no provision had been made to save those who should venture to transgress his law. He claimed that angels needed no law; but should be left free to follow their own will, which would ever guide them right; that law was a restriction of their liberty, and that to abolish law was one great object of his standing as he did. The condition of the angels, he thought, needed improvement. Not so the mind of God, who had made the laws and exalted them equal to himself. The happiness of the angelic host consisted in their perfect obedience to the law. Each had his special work assigned him; and until Satan (Lucifer) rebelled, there had been perfect order and harmonious action in heaven.' It is reported in the Scriptures that 'there was war in heaven. Michael and his angels fought against the dragon; and the dragon fought and his angels, and prevailed not; neither was their place found anymore in heaven...' and Ellen G. White says, 'No taint of rebellion was left in heaven. All was again peaceful and harmonious as before,' (GC-23).

"The great controversy, becoming now an open war between good and evil, light and darkness, didn't involve only the two powers in contention, but was also extended to the whole creation of God. The arch rebel who had tried unsuccessfully to win the sympathy of the entire universe to his cause, finally succeeded in his attempt to deceive Adam and Eve, the representatives of our race in the assembly of all the unfallen worlds that comprise the whole creation of God. That's why Lucifer found a shelter in our world having in addition taken, by usurpation, the place that once belonged to Adam, the father of all humankind. Job 1:6,7 reports one of these assemblies when the sons of God came to present themselves before Him; and Satan (Lucifer) being found among them and asked whence he came, answered: 'From going to and fro in the earth, and walking up and down in it.' Satan then, as he always does, made some insinuations about Job's character and questioned his faithfulness to God. Whoever takes a stand on the side of the Lord will always have his antipathy and opposition. We can observe that all along the church's history. Even when human race was just beginning in the planet, Lucifer started his evil plans of corruption and destruction, trying to concentrate them in walled cities so he could more easily work on the social, economical and moral aspects of human society in order to lead them into a deeper level of rebellion against God and His laws, making sure that their physical, moral and spiritual resemblance with the creator would be deformed in such a way that man would no longer reflect the image of God.

"The plot involving the dinosaurs and his very elaborate plan to prevent human beings from fulfilling the creator's purpose of peopling the planet, subduing it, and ruling over it in subjection to His authority was only a tiny part of Lucifer's plan of action; and when the flood was sent, even though the condemned generation of men and all the dinosaurs perished in the waters, his plans still went on with the subsequent generations after Noah. Today, enlightened by the Spirit of God, we can discern his actions in the play and counter play of nations; the way he acts, influencing those who are in a position to make important decisions in the government level making it possible that laws that would favor those who act on his will are approved so God's people will meet very hard opposition as they try to be obedient to their Lord and Master.

"Parallel to all his attempts to destroy humankind by wars, terrorist attacks, and contamination of the air causing the spreading out of lethal diseases; he's given special attention to his work of deception earning his reputation as the master of deceit. Now, the dinosaurs, being used in a different way they were used in the past, are still a powerful instrument in the hands of the enemy to create the false idea that life started on our planet as a result of millions of years of evolution, denying the existence of a personal God who created and maintains every thing working for the good of his creatures.

"The theory of evolution, the spiritualism, and the preaching of *salvation by grace* – cheap grace – that ignores the sanctity and the validity, in this end of time, of God's commandments as they are expressed in the Decalogue, will play an important role in Lucifer's work of deceiving the multitudes and leading them far away from God's way of life. And as time rushes its flight toward the end, the character of God, upon which Lucifer has cast the most unfair accusations, is being evaluated by the expectant non-fallen beings in God's entire creation; there coming, finally, a time when it will be vindicated before all the intelligent creatures that inhabit the universe; as it is stated by John, the prophet of Patmos, in the book of Revelation: "Great and marvelous are thy works, Lord God Almighty; just and true are thy ways, thou King of saints. Who shall not fear thee, o Lord, and glorify thy name? For thou only art holy; for all nations shall come and worship before thee; for thy judgments are made manifest. Rev. 15:3, 4

Appendix 1
Brain Size and Intelligence

Humans have brain sizes ranging from up to 2000 cc. with the average being about 1400cc.

A larger brain does not necessarily imply higher intelligence. This means that someone with a larger brain may be less intelligent than someone with smaller brain; however, a larger brain correlates with higher intelligence, particularly between different species; this means that species with larger brains tend to be more intelligent creatures. The sheer size of the brain is relevant for two reasons: Most obviously, a small brain cannot hold as many brain cells as a large one. Less obvious, but more important, is that the true quality of a brain must be measured by the complexity of linkages between cells.

Besides the size, there are many other factors that affect the intelligence of a brain, such as the density of the brain, its structure, the fissure and the size of the frontal lobe relatively to the rest of it. The surface of the cerebral cortex has a lot to do with intelligence. The more convoluted the surface of the brain is, the more cerebral cortex a person has and therefore, a higher chance of being more intelligent.

The theropods, comprising a group that includes dinosaurs such as Allosaurus, Tyrannosaurus rex, and Coelurosaur Troodon, are believed to be the smartest dinosaurs, if the ratio between body size and brain size (a rough measurement of intelligence) is considered. Troodon had one of the largest brains relative to its body size of any dinosaur. Being a close relative to Velociraptor, and having many characteristics in common with Deinonychus, this dinosaur was a meat-eater the size of a man, having a brain as big as an avocado pit. It was not only the smartest dinosaur, but the smartest animal of dinosaur times. Its method of locomotion was bipedal, similar to the method used by today's famous bird of the southwest – the roadrunner. Troodon's rear leg bones and other features indicate that it was also swift running; its large eyes set partially forward in its head indicated that it probably had binocular vision for depth perception. Its arsenal also included flexible grasping front hands with sharp recurved claws. It may have been a stalker, or a fast pursuer, or perhaps both; but regardless, it was an efficient two-legged predator well adapted to capturing elusive prey.

The biggest brained dinosaur of all was probably Tyrannosaurus rex, since it was such a huge animal. Its brain was about as big as ours, but it was many times bigger than we are. Stegosaurus was a tiny-brained dinosaur compared to its size. Its brain wasn't much bigger than a ping-pong ball and its body was the size of a truck.

Appendix 2
Plant-Eating Dinosaurs and Their Impact on Nature

Charles Darwin in his book *The Origin of Species* explains about the struggle for existence, which is closely related to the high rate at which all living things tend to multiply. He mentions the doctrine of Malthus, according to which "Although some species may be now increasing more or less rapidly in numbers, all cannot do so, for the world would not hold them. There is no exception to the rule that every organic being naturally increases at so high a rate that if not destroyed the earth would soon be covered by the progeny of a single pair." In addition, Darwin mentions an important fact related to the elephants reproduction. He says, "The elephant is reckoned to be the slowest breeder of all known animals, and I have taken some pains to estimate its probable minimum rate of natural increase. It will be under the mark to assume that it breeds when thirty years old, and goes on breeding till ninety years old, bringing forth three pair of young in this interval; if this be so, at the end of the fifth century there would be alive fifteen million elephants, descended from the first pair. On the other hand, in some cases, as with the elephant and rhinoceros, none are destroyed by beasts of prey, and even the tiger in India most rarely dares to attack a young elephant protected by its dam."

Organisms that lay eggs in a large number, protecting them until they hatch; and then protect their young until they can take care of themselves, are the ones which will greatly multiply in number, so he concludes by saying that "In all cases, the average number of any animal or plant depends directly on the number of its eggs or seeds."

Now, if we consider that the small-and-medium sized plant-eating dinosaurs laid eggs in a great number, and that some of them lay on their nests hatching their eggs, and later fed and protected their offspring from predators, we can understand why they grew to such a high number in a relatively short period of time (nearly 2000 years until the flood). With the giant-and-super-giant dinosaurs such as Apatosaurus, Diplodocus, Camarasaurus, Seismosaurus, Brachiosaurus, and Ultrasaurus, the situation was not so different, since they, too, multiplied considerably reaching a very great number. These gigantic plant-eating dinosaurs, as the elephants, after reaching adulthood; and consequently, having grown to a huge size, scarcely could be vulnerable to predators; with some of them living longer than human beings do today, and laying about 100 eggs in each nesting season.

According to some specialists, the super-giant plant-eating dinosaurs would have to eat every waking moment to get enough food to keep their large bodies alive (an elephant spends, on average, 16 hours a day eating). These sauropods seemed to have every adaptation for continuous eating, including having nostrils on the top of their heads so breathing would not interfere with eating. They did not chew the vegetation they ate, and were equipped with peg-like teeth in their front jaw to enable them to chip off leaves, swallowing up a large amount of them in a very short time. They also had a gizzard to help them in the digestion of the toughest plants they could eat.

Consider some important facts about the plant-eating dinosaurs:

- They lived in colonies.

- They traveled long distances searching for food and nesting grounds.

- They ate a great deal of food daily (some ate a ton of foliage a day)

- The medium sized plant-eating dinosaurs such as Iguanodon which grew to 20feet long. Hadrosaurus which grew to 43feet long, and the Ceratopians, were all very abundant and extremely efficient herbivores.

- Some of them, especially the small-sized dinosaurs, were able to thrive even in severe drought and scarcity of food.

- The much larger dinosaurs were very heavy, and their treading on the ground would have made it packed and hard, making it very difficult for plants to grow back.

- If after they had passed, eating all the vegetation, it still grew back; it would require some time until the recovery could take place; and by then, the climate (frequency of rain, temperature, and other factors) would have suffered changes that in some places could have been irreversible.

Though it is possible that, in the antediluvian world, nature was well capable of recovering, even from a massive attack represented by the intensive grazing and browsing by those huge beasts, we must consider that if even the most luxuriant vegetation, as it was the case before the flood, is consistently and intensively abused, through over grazing and browsing during the critical growth periods, the plants lose vigor, their roots begin to shrink, and they eventually die, being replaced by an inferior species. In the meantime, the soil, no longer adequately protected by a plant cover or firmly held by a thick mass of roots, begins to wash away. Thus, moisture and nutrients needed by the plant are lost. Unless corrected, the process of deterioration will continue faster and faster.

The plant-eating dinosaurs existed in myriads of them, there being strong evidences that they roamed in the North American plains in herds as numerous as the large herds of buffalo seen in this country in the past. In other continents, the situation was not different, based on the number of fossils found everywhere. If they did not have the capability to effect a total devastation of the vegetation wherever they went, they at least had a great potential for massive damage of the flora on a worldwide scale; fulfilling, in this way, their part in the rebellious plan of their developer: to promote ecological unbalance and the creation of deserts everywhere.

Appendix 3
Herrerasaurus: A Proto-Dinosaur

In the 1970s in Argentina, at the farm of Victor Herrera, scientists found a dinosaur fossil, which they named *Herrerasaurus*, after Mr. Herrera. The dinosaur specimen, 4 meters (13feet) long, had characteristics of both Saurischians and Ornithischians. Later it was noticed its resemblance to another dinosaur fossil named Staurikosaurus discovered in Brazil, back in the 1960s. More recently (1993) another herrerasaurus-like fossil was found in the same general area, and was named Eoraptor, or "dawn thief," which appeared to be closely related to Herrerasaurus, but smaller in size and slightly older. Both Eoraptor and Herrerasaurus seem to have been small, medium-sized carnivores.

These curious animals have some basic theropod characteristics, but lack others; in fact, they lack some dinosaurian characteristics as well. Current thinking has this dinosaur branching off earlier than the split between Saurischians and Ornithischians, but after dinosaurs separated from other reptiles. Herrerasaurus, they say, is one of two primitive animals that mark the transition to dinosaurs. The Herrerasauridae and Eoraptor may be the earliest group of theropods, or it is quite possible that they are not even theropods at all, but rather non-dinosaur "dinosauromorphs" closely related to the ancestor of dinosaurs; though other specialists still see in them a physical resemblance and posture close to the later Allosaurus, Tyrannosaurus rex, and other theropods.

According to some scientists, these Herrerasaurus had five-fingered hands, digit IV greatly reduced and digit V being barely there. Still, according to the same scientists, Coelophysis seems to represent the next step along the way to a tridactyl or three-fingered hand. Digit IV is reduced and digit V is gone. By the time Allosaurus comes along, only digits I, II and III remain.

In 1999, Wagner and Gauthier proposed the Frame Shift Hypothesis. Nathan Smith modified and greatly elaborated on this idea in a paper presented at the October 2003 meeting of the Society of Vertebrate Paleontology. These scientists pointed out that the genetic and chemical signals that cause the digits to form are distinct from the genetic and chemical signals that cause each digit to develop in a particular way. Smith proposes that some time after the stage represented by Herrerasaurus, theropods lost digit I.

The Plot's Theory

The Plot's main argument is that Lucifer, as a possessor of the knowledge of DNA code, tried to manipulate it in order to turn on the genes that would codify for the development of five fingers, probably in an attempt to develop in those beasts, the characteristic that distinguishes humans from other animal species giving them the capability of handling their prey with the same degree of sophistication found in human beings – in reference to the use of their hands. It seems that, after trying for a while, the developer of those proto-dinosaurs found it easier to develop their successor dinosaurs with only three fingers giving up on the idea of five fingers for carnivorous dinosaurs.

Appendix 4
Dinosaurs Had the Best of Birds

The longstanding debate among scientists about the origin of birds is a very fierce one. While there are those who advocate the idea that these avian, warm-blooded, egg-laying vertebrates evolved from dinosaurs; there are also the ones who argue in favor of a different ancestor for them. And though not agreeing on the particular point of the origin of birds, these men of science are in perfect agreement regarding the other aspects of the evolutionary theory that relates the origin of life in our planet to an evolutionary process that had its beginning with bacteria and mollusks, millions of years ago, evolving to what we see today.

In fact, as far as dinosaurs and birds are concerned, what we observe is the presence of a plethora of anatomical and skeletal resemblances between the two species (theropods and birds), which are as follows:

- Birds and dinosaurs lay eggs.

- The soft anatomy (musculature, brain, heart, and other organs) all is similar.

- Among others, there is the resemblance on the toes and fingers that always ended in sharp and curved claws, like birds claws; the four toed foot supported by three main toes; the similar eggshell microstructure; and the clavicles (collar bone) fused to form a *furcula* (wishbone).

- Predatory dinosaurs had a bird-like pulmonary system.

Though some scientists have proposed that predatory dinosaurs had lungs similar to those of crocodiles and other reptiles, a new study published in the journal *Nature* (July, 2005, pp 21-27) suggests that the ancient beasts boasted a much bigger and more complex system of air sacs similar to that of today's birds.

The pulmonary system of meat-eating dinosaurs such as Tyrannosaurus rex, in fact, shares many structural similarities with that of modern birds, which, from an engineering point of view, may possess the most efficient respiratory system of any living vertebrate inhabiting the planet.

The resemblances between birds and dinosaurs now assume a curious character as the scientists strive to find a dinosaur ancestor to link to Archaeopteryx and then link Archaeopteryx to modern birds.

This is where things get really complicated, because in the last decade, several significant dinosaurs with bird-like features and "primitive birds", they say, with dinosaur-like features have been found around the world, mainly in China leading the scientists to a difficult dilemma. The closest dinosaurian relatives to birds, say some evolutionists, occur in the fossil record after Archaeopteryx. Unless Velociraptors and their kin perfected time travel, there is no way they can be the ancestors of a bird that lived sixty million years earlier.

Though this seemingly unsolvable puzzle appears to have no simple viable solution, *The Plot*, in the person of its main character, Richard Campbell, boldly suggests a way out of this perplexing situation: "Birds didn't evolve from dinosaurs, but in dinosaurs those features which were so distinctive in birds, were methodically developed by Lucifer's powerful mind using the DNA code manipulation. And we know that he could go much further than our most brilliant scientists can go today, due to his superior mind – a powerful genius that no human being can ever match.

Appendix 5

Dinosaurs, Ellen G. White, and the Bible

> *In regards to creatures that lived before the flood and were not created by God being consequently left out of the ark, Ellen G. White wrote, "Every species of animals which God had created was preserved in the ark. The confused species which God did not create, which were the result of amalgamation, were destroyed by the flood..." (Spiritual Gifts, page 75).*

The term "genetics", coined in 1909 by British biologist William Bateson and other related terminology coined by later scientists did not exist in 1864, when Mrs. White wrote about the flood that occurred in Noah's time. Therefore, there's a great probability that she had never heard about words such as *genes, genetics, clone, genetic engineering,* and *DNA code*; because when this branch of science came about she had already died, having passed away in 1915. Consequently, she could only have spoken about the mixture of different species using the contemporary terminology used in the latter part of her lifetime. Charles Darwin himself never used the word "amalgamation" in the course of explaining his theory in 1859. Since he spoke about natural selection involving the transmission of characters through heredity explaining that under the influence of life conditions and environmental factors, variations were generated among the species, favoring some individuals to the detriment of others determining by this the survival of the fittest.

What we understand from Ellen G. White's quote above is that Satan had been tampering with nature by making amalgamations, or mixtures among men and among beasts. Obviously, we believe that when she wrote on this matter, she was inspired; though we also believe that, as any other prophet of old; she might have spoken about subjects and things that she did not fully understand because they were aspects of the truth reserved to be understood in our time, not in her time. Moreover, if the word amalgamation in Ellen G. White's statement above means the breeding of two different species in order to obtain a third one, in no dictionaries researched by the author has amalgamation been defined in this sense. Not all of them will blend except in a very few cases with the resultant offspring being sterile, and not able to perpetuate itself by generation of its own offspring. Based on that, it is difficult to see how Lucifer could have mixed up different species such as reptiles, mammals, and birds without resorting to the knowledge of DNA and clever manipulations of its code, since this mixture of species could only be done on the DNA level.

Mrs. White pictures Satan as stalking over the earth, bent on disorder and *devastation*, attributing to him the developing of confused species of animals. In another reference to amalgamation she clearly attaches him to the cause of certain changes occurred in the flora and fauna of our world. She states, "In the parable of the sower the question was asked the Master, 'Didst not thou sow good seed in thy field? How then hath it tares?' The Master answered, 'An enemy hath done this.' All tares are sown by the evil one. Every noxious herb is of his sowing, and by his ingenious

methods of amalgamation, he has corrupted the earth with tares."
(*Selected Messages*, book 2, page 228.)

The Plot is a book that presents a new idea about the origin
and demise of the dinosaurs, founded on two basic reasons. First,
science has been unable, so far, to present convincing evidence
that would, once and for all, put an end to the long-standing
debate involving the origin of life in our planet. The second reason relates to the very nature of those huge creatures whose anatomy, warm-bloodedness and cold-bloodedness cleverly combined,
reveal a different standard of animal species, apart from the whole
plan of creation giving the impression that their developer was
trying to equip them with the best of reptiles, mammals, and birds
at the same time making them capable of thriving in the most hos-
tile environment; and becoming, if possible, the dominant species
above all others.

In the Bible, we find references to huge animals, *Behemoth*
and *Leviathan*, which some erroneously refer to as dinosaurs. Job
40:15-24; Job 41:1-34.

The name *Behemoth* is a transliteration of the Hebrew. It is
the plural form of the common Hebrew word *behemah*, translated
as "cattle". (*SDABC*-606) Therefore, the animal here described
(a cattle-like animal) seems most likely to be a mammal and
not a mixture of reptile, mammal and bird, as paleontology has
described the dinosaurs. Other characteristics of the animal, such
as the strength of its limbs, the "thick and muscular tail" and the
sharp teeth like a "sword", made some authorities in the subject to
identify *Behemoth* with the hippopotamus of our time.

Leviathan, the other huge creature mentioned in the Bible,
which some Bible commentators admit most likely to be a croco-
dile, based on the similarity found in the Leviathan's description
by the Bible, and this huge reptile of modern days is described in
the following terms in The Seventh-day Adventist Bible Commen-
tary: "The creature is represented as wild, fierce, and ungovern-
able. Having a mouth of large size and armed with a formidable
array of teeth. The body is covered with scales set near together,
like a coat of mail." (*SDABC*-607) In addition, others even argue
that the Leviathan could be a Plesiosaur, a huge aquatic animal

that lived in dinosaurs' time, saying that supposedly, Plesiosaurs could breathe fire. But in any case, we should consider that dinosaurs, according to their definition by paleontologists and naturalists, were land animals, not sea-monsters.

Therefore, the Bible, except in the text of the parable of the sower found in Matthew 13:24-30, which is a reference to Satan's power to interfere with nature, does not say anything that can be understood as an explicit reference to dinosaurs.